A Roller Coaster Ride through Relativity

J. Oliver Linton

ISBN 978-1-4466-6152-9

Contents

Preface

This book is written for anyone who is doing or has done A level Physics and/or A level Maths and the intention is to explain the theories of Special and General Relativity as far as is possible using concepts and language appropriate to that level of knowledge.

Many excellent books on relativity attempt to take the general reader all the way from basic mechanics to tensor theory in one broad sweep or try to explain black holes, for example, without using a single equation but it seems to me that these books, excellent though they are, in the end satisfy neither the general reader nor the expert. There are a few books which really do attempt to justify what they claim in appropriate mathematical language but often they try to go a bit too far and are a bit too rigorous - understandably, as they are mostly written for first year university students. There are also, regrettably, quite a few books out there which are either over-simple or frankly, wrong. As Einstein himself said, 'you should explain your theories as simply as possible - but not more so.'

This book is, therefore, an attempt to fill the gap by explaining Einstein's theories using plenty of simple algebra and numerical examples but without introducing any mathematical techniques beyond some very simple calculus.

There is a story about Sir Arthur Eddington who wrote many books explaining and popularising Einstein's theories. A journalist once exclaimed to him that he was one of only three people in the world who really understood the General Theory of Relativity. Eddington was silent. When asked why he did not say anything he replied "I was just wondering who the third person might be."

In the century that has passed since Einstein's first publication the scientific world has gradually come to an acceptance that the General Theory of Relativity – at least on a large scale – is the way the world works and its predictions have now been verified countless times to an amazingly high degree of accuracy. But as to the number of people who really understand it, you can probably still list their names on a single sheet of paper – and you won't find my name on the list! On the other hand, the basic ideas of Relativity and some of its bizarre consequences including the existence of Black Holes have become common knowledge and anybody with an inquiring mind will want to know something about how these claims are justified.

So how much is it possible to achieve using just A level ideas? The answer is quite a lot. The fundamental theorem of special relativity (time dilation) requires nothing more than Pythagoras' theorem and the remaining theorems require as much mental agility as algebraic competence. All the proofs given have been chosen carefully and the proofs of the 1g rocket problem and the addition of velocities are new in the sense that I have not seen them proved this way elsewhere. The approach to the famous equation $E = mc^2$ is also slightly novel and avoids the complications of quantum theory which is a very unsatisfactory feature of any proof which relies on the behaviour of a photon in a box. At each stage in the argument I have been careful to show how the relativistic expressions reduce to Newtonian ones when v is much less than c.

When it comes to General Relativity, one can really only get as far as Einstein's 1911 paper in which he describes the bending of light and gravitational time dilation. The extra effects due to the distortion of space and time revealed in the full theory which appeared in 1917 lie beyond my competence. Nevertheless, it is possible to calculate some of the predicted properties of those extraordinary objects known as black holes and even to start to discuss the numerical properties of the universe as a whole.

A quick flip through the book will reveal that it is full of equations and if it is true that every equation reduces one's potential readership by 50% then my readership is going to be vanishingly small! I sincerely hope that this is not the case, however. Galileo said the the Book of Nature is written in the language of Mathematics and if we are to understand the world we live in, we must accept the fact and continue to practice our facility with that language even when we have left our formal mathematical education behind. Besides – equations have an important property which, like other theorems which occur in this book, is so important I shall put it into a colourful box:

> ### The Fundamental Principle of Reading Mathematical Books
>
> *All mathematical equations can be admired or ignored as required. All that you need is confidence in their veracity.*

A mathematical proof is like the Title Deed to a house. It is very important that the Deed exists and that it is kept somewhere safe. When you first buy a house, you might be curious enough to glance through it to see what it says but you are unlikely to understand much of the legal language it is written in; nevertheless, you have employed a good solicitor to make sure that it is in order and you can, at least, admire the fancy paper it is written on.

I urge you to regard the equations and proofs in this book in the same way. Have a glance through them; try to get to grips with a few of them but don't think you have to understand every equation – just sit back and admire them. Do take the trouble to get your calculator out and verify some of the figures though otherwise you may find them difficult to believe. Then when you have finished the book you can put it in a safe place in your bedroom and go to sleep in the comfortable knowledge that even if you still can't really believe that clocks in motion go slow and that penny's bend when they accelerate, the proofs are quietly sitting there on your shelf so it must be true after all.

Prologue to Chapter 1 - The initial climb up

Let me take you on a roller coaster ride through Relativity. You will see many strange sights along the ride and hear many strange stories - many of which you will find hard to believe. But at the end of the ride, you will be able to look at the world about you and the stars above with a new and altogether deeper understanding. Have you got your ticket? Then let's get going . . .

Together we climb into the train and sit down.

'There's not much to hang on to', you say.

No, you're right there. In front of us there is absolutely nothing except what looks like a single gear stick with a black knob on the end. On the knob there is some writing engraved in white which says simply: 'The Fundamental Principle of Special Relativity'.

That's all there is, I am afraid, and you have to hang on to it like glue. If ever you let go of this Principle, you are lost. The Principle itself is seemingly innocuous, almost self evident, and yet Einstein showed that it leads to an amazing series of almost incredible consequences. While the roller coaster climbs to the top of the first hill, let me tell you what this wonderful Principle is

> ### *The Fundamental Principle of Special Relativity*
>
> *The laws of Physics are identical for all observers in relative (uniform) motion with respect to each other*

'Is that it?' you ask. *'I thought that was obvious'*.

Well, yes, it is. After all, when you pour a cup of tea from a tea pot into a cup, it doesn't matter if you, the cup and the teapot are all hurtling down a (straight) railway line at a (constant) speed of 125 mph in a railway carriage. Nor does it matter that the whole train and indeed the whole Earth is hurtling round the sun at an (almost constant) speed of 30 km s^{-1}, nor does it matter that the whole solar system is hurtling round the galaxy even faster than that! The laws of physics which govern the way the tea falls are just the same. Nor would you expect calculators to give different answers or musical instruments to make different sounds just because they were moving. Surely all the laws of physics are the same whether you are moving or not.

We can restate our Principle in an equivalent form like this:

> *It is impossible to carry out any experiment inside a closed laboratory which will detect whether or not the laboratory is moving. Absolute motion is meaningless, only relative motion can be measured.*

1

Sounds very plausible, doesn't it?

'Sounds perfectly obvious to me.'

I agree. But there is, perhaps, one way in which you just might be able to tell if you were moving or not. What if you were to measure the speed of light travelling in different directions? Suppose that you discovered that the speed of light measured in one direction was greater than the speed of light travelling in the opposite direction? What would you infer then? Surely it would be reasonable to suppose that your laboratory was in fact moving through space and in one direction the speed of light and the speed of the lab were adding together while in the other direction the two speeds were subtracting.

It is actually very difficult to do this experiment because you have to measure the speed of light very accurately but eventually, as we shall see later, a suitable experiment was performed by two physicists called Michelson and Morley, but the results were disappointing. The speed of light seemed to be constant in all directions.

While most scientists tried to explain this result away by means of various subterfuges, Albert Einstein merely accepted it as a necessary consequence of the Fundamental Principle namely:

The speed of light in a vacuum is a universal constant and will always be the same even when measured by different observers in relative motion.

(It is worth noting that Einstein was not aware of the results of the Michelson-Morley experiment when he first worked on the theory of Special Relativity. Einstein - like Galileo, Newton and Maxwell - was one of those geniuses who did not really need to rely on experimental evidence to point the way forward. He just knew his theory was right!)

We have now cranked our way to the top of the first hill. Let's pause a minute and take a last look around at our cosily familiar world. Far below you can see your twin brother thumbing through what looks like a Thomson travel brochure. Away to your right you can see some workmen doing some maintenance on a large clock which towers over the rifle range. The shrieks of the children riding the Ghost Train catches your attention and you watch for a moment as the long train rushes into a tunnel. You notice the rear of the train vanish into the tunnel at the precise instant that the engine begins to emerge from the other end. Down below you notice some children playing shove-halfpenny on a cracked old table and two boys tossing a couple of footballs. In the distance you can see a fast flowing river. Wait a minute - a race is about to begin...

The River Race

Two friends, Albert and Beatrice if you like, have agreed to a rowing race on a fast flowing river. Albert is going to row 100m directly across the river and back again. He knows that in order to proceed at right angles to the flow he will have to aim upriver a bit and this will slow him down on both halves on the race, but he doesn't think he will be slowed down too much. Beatrice is planning to row to point B 100 m upstream and back again. She knows that it will be hard work rowing up stream but she reckons that she will be assisted just as much on the way home as she is hindered on the way out and that she will win the race overall.

Who do you think will win the race?

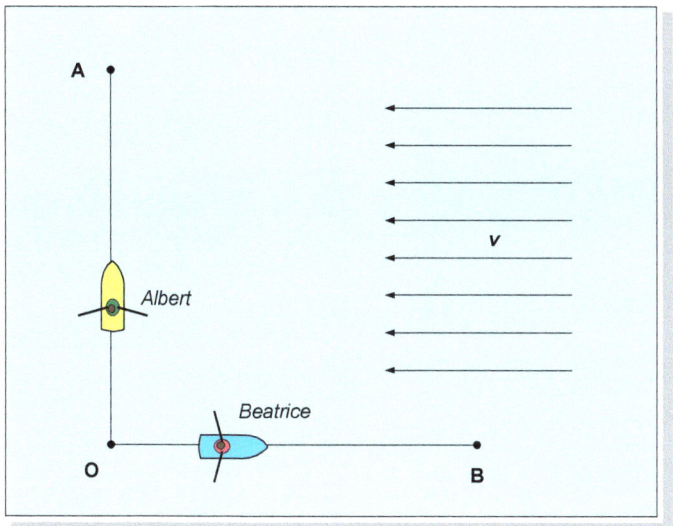

You watch. As expected, Albert makes steady progress across the river and easily beats Beatrice, who is pulling hard upstream, to the first turn around point. But when Beatrice eventually reaches her turn-around the excitement and the cheering begins to mount because the current is rapidly carrying her back to the start while poor Albert is still struggling across the river. Nevertheless, Albert has gained too much of a head start and in spite of the assistance of the current, Beatrice cannot make up the lost ground and loses the race.

Why is this? Let's work it out using some typical figures.

If we assume that the river is flowing at 1 ms^{-1} and that both rowers row at 2 ms^{-1} through the water, it is easy to use Pythagoras' theorem to see that Albert will move across the river at a resultant speed of $\sqrt{2^2 - 1^2} = \sqrt{3}$ ms^{-1} and that he will complete the race in 115 s.

Beatrice, on the other hand takes 100 s to get to the turning point (travelling at an effective speed of 2 - 1 = 1 ms⁻¹) and in spite of the assistance of the stream, she cannot make it back in time to win the race because even at a speed of 2 + 1 = 3 ms⁻¹, the return journey takes her 33s.

The difference in journey times is due to the different geometries of the two paths and of course it depends on the speed of the river. It is worth pointing out that if the river was flowing at 2 ms⁻¹, the same speed that Albert and Beatrice can row, neither rower would complete the race at all but it turns out that at speeds less than 2 ms⁻¹, Albert always wins.

What is the significance of this pretty little story? Well in 1887 a famous attempt to measure the speed of the Earth through the supposed æther was made by Michelson and Morley using a device called an interferometer. The apparatus looked (schematically) something like this:

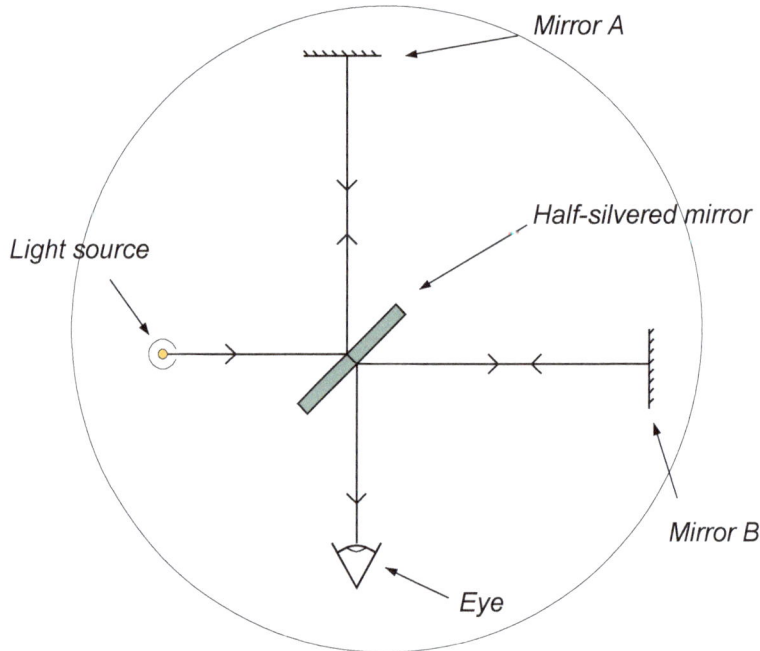

Light from a monochromatic light source is split by a half silvered mirror into two beams which travel out to two distant mirrors A and B (just like the two rowers Albert and Beatrice). When they return, the same half silvered mirror recombines the two beams into one again. If the two arms are of exactly equal lengths (and if the Earth is stationary) the two light beams will take exactly the same time to get to the mirror and back and will therefore arrive back exactly in phase and will interfere constructively (i.e. they will produce a strong interference fringe). If the Earth is moving, however, according to the æther theory of light, the beam which is travelling parallel to the direction of the Earth's motion through space will be delayed with respect to the other beam and the fringe pattern will shift, perhaps showing destructive interference instead of constructive interference. In order to eliminate the effect due to the inevitable slight difference in arm lengths, the whole apparatus was designed to be rotated slowly through 90° at a time.

Now the Earth moves round the Sun at about 30 km s^{-1} - that is about 0.01% of the speed of light. This makes the calculations a bit difficult because the important figures occur in the 10th decimal place and your calculator may not be accurate enough - but the result nonetheless is that Michelson and Morley expected to see a fringe shift of about ± half a wavelength. Not much, but easily detectable all the same.

Well, what did they see? They looked for a whole year and found precisely - nothing. No fringe shift at all. Whichever way you looked at it the experiment was either a triumphant failure or a dismal success - for it seemed to indicate that Earth wasn't moving at all!

Various theories were put forward to explain the result. It seemed incredible that the Earth was the only thing in the Universe with zero speed, so perhaps the Earth 'dragged' the æther around with it, or perhaps the arms of the apparatus changed in length according to how they were moving.

None of these are correct. The truth of the matter is, of course, that the result of the experiment does not need explaining! It is simply a fundamental fact about the universe we live in. The speed of light in a vacuum is a fundamental constant and will always be the same even when measured by observers in relative motion.

'I don't have a problem with that.'

You will.

'Why?'

The problems all arise when you consider measurement made of the speed of light by two different observers who are in relative motion. The first thing you will have to accept is:

Bizarre consequence number 1

Moving clocks run slow

The roller coaster gives a violent jerk and suddenly you are accelerating rapidly down the long descent. 'Hang on to that Principle' I shout as your arms flail around wildly trying to find something to grab hold of. As we gather speed I glance at the clock tower and note with satisfaction that the hands of the clock are not moving quite as fast as they were...

Chapter 1 - Time Dilation

The simplest possible clock is a beam of light bouncing backwards and forwards between two parallel mirrors. If the mirrors are separated by a distance l, the time for each 'tick' (i.e. there and back) is $t = 2l/c$ (Of course you remember *that speed = distance over time* and that *time* is *distance over speed*) It would be perfectly possible for an electronic circuit to count the number of ticks and provide a suitable display in hours, minutes and seconds. (I do not know if anyone has ever made such a clock. If the arm of the clock was 1 metre long, it would 'tick' at a rate of 6.7 GHz. This is not a lot faster than the clock speed of your average PC.)

Now suppose you are watching such a light clock passing by in a spaceship travelling at a sizeable fraction of the speed of light v. (Of course you can't see the light beam but you can imagine it!) The arm of the clock is at right angles to the direction of motion of the ship so you have no difficulty in verifying that the arm has a length l as the ship goes by. On the other hand, you can see (imagine!) the light beam moving in a diagonal line like this:

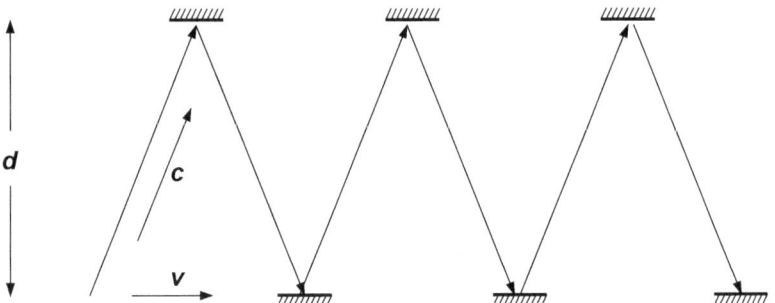

Now because of the Fundamental Principle, you see the light beam moving at the normal speed of light. This means that it is going to take longer for it to travel the distance l and back again just like Albert on the river (albeit for a slightly different reason). In fact the effective velocity across the ship is $\sqrt{c^2 - v^2}$ and the time taken for the return trip is $\quad t = \dfrac{2l}{\sqrt{c^2 - v^2}}$

This means that the clocks on board the ship (and everything else as well) appears to me to run *slow* by a factor of $\quad \dfrac{c}{\sqrt{c^2 - v^2}} \quad$ or, as it is more usually written

$\dfrac{1}{\sqrt{1 - v^2/c^2}}$. This factor is often called γ (the greek letter gamma) and is always greater than one. It rises to ∞ – as v gets closer and closer towards c.

We can summarise what we have derived so far as follows. If it takes T_0 seconds to boil an egg, it will appear to me that the eggs in the spaceship take T seconds where:

$$T = \gamma T_0 = \frac{1}{\sqrt{1 - v^2/c^2}} T_0$$

Let's put some figures into the formula and see what we get:

v (as a % of c)	γ
50	1.15
60	1.25
70	1.40
80	1.67
90	2.29
95	3.20
96	3.57
97	4.11
98	5.03
99	7.09

If we plot a graph of these figures we get this:

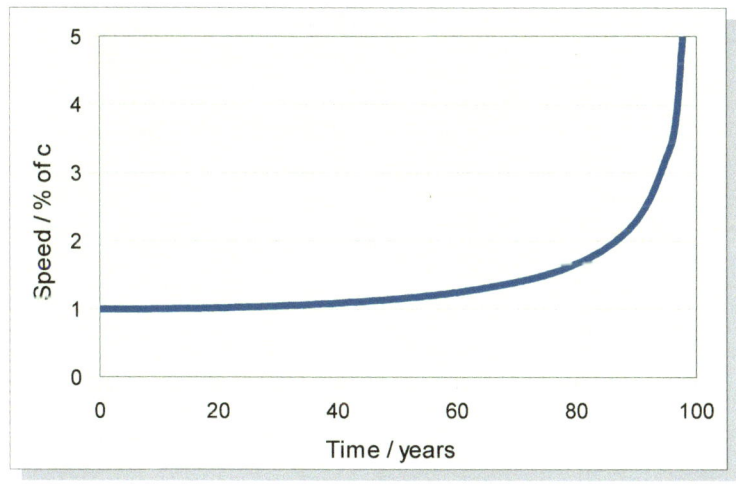

What this means is that if you travel at 50% of the speed of light, your clocks go 15% slower than clocks which are stationary. Travel at 80% of the speed of light and your clocks are 67% slower (ie at 3/5 of the rate of a stationary clock). Travel at 99% of the speed of light and every second of your time takes over 7 seconds of 'stationary clock' time!

'That's just ridiculous!' You say.

'I mean, what would it be like to live in a world in which all the clocks went 7 times slower than normal? It would be a crazy world! Everything would appear to be slowed down! Cars would crawl along at 10 miles per hour. A game of football would take all day! A falling stone would appear to fall like a feather!'

Hang on a minute, you are not thinking quite straight. If the clocks go 7 times slower, ...

'Oh! I see what you mean': you interrupt. *'If the clocks go slower, stones will actually take 7 times fewer seconds to fall so everything will actually look as if it is happening faster! Is that it?'*

No, no, that's not right either! The point is that it is not just clocks which go slow - everything goes slow, you might say that time itself goes slow. When you watch a falling stone and time it with a stop watch, the stone falls slowly, the clock ticks slowly and your own mental processes think slowly as well. When the stone reaches the ground, the clock reads exactly what you expect it to read and the process seems to have taken exactly the same time as usual. In short, the whole process looks perfectly normal to you. In fact it doesn't just *look* perfectly normal to you, it *is* perfectly normal to you. You must remember that time is only dilated (ie stretched) from the point of view of someone else who is moving with respect to you. In your moving spaceship, everything looks normal to you. It is only me, on stationary Earth who sees your clocks going slow and your stones falling like feathers and your eggs which take ages to boil!

After a moment's thought you exclaim: *'That can't be right! and what's more, I can prove it!'*

Go on then.

'Well you say that the clocks in the space ship appear to go slow because the spaceship is moving and you are stationary. But from my point of view in the spaceship, it would appear that it was the clocks on Earth which were going slow because, as you said yourself, all velocities are relative and you can't tell who is actually moving.'

Bravo! You really are beginning to think like a relativist!

'What do you mean, Bravo!? Haven't I just disproved your theory?'

Well no, you haven't. What is wrong with both clocks appearing to go slow?

'Surely that's obvious. All you have got to do is put the clocks side by side and see which one is going slow!'

But how are you going to do that?

'Well, as I go past you we will synchronise our clocks, and then a short while later we will see which clock has gone slow.'

But by that time you will be hundreds of miles away.

'Well, I'll send you a radio message with a time signal'

But we will disagree about how long it takes the radio signal to get back to me.

'Well - I will have to stop the spaceship and turn it round and..'

Aha! You are going to STOP the spaceship! As soon as you do that, the motion of the ship stops being uniform and we must examine very carefully what happens when the motion of the clock changes. The slowing down of the ship introduces an asymmetry into the situation which means that when you bring the clocks back together, the clock that was in the spaceship will indeed be found to be slow compared to the clock which remained back on Earth.

'Are you sure?'

Sure I'm sure. Believe it or not, this experiment has actually been done using extremely accurate atomic clocks and the results confirm Einstein's theories exactly. Mind you, the effect is incredibly small. Suppose you synchronise two clocks on the ground and then fly one of them round in a jet plane at a speed of 200 ms^{-1} for say 10 hours. What will the expected time difference due to special relativity be? (I must say special relativity, because in the real experiment the effects of general relativity have to be taken into account as well.)

The first thing we have got to do is to work out y for a speed of 200 ms^{-1}. If you try doing this on a calculator you will run into a problem. The velocity of light is very large. It is in fact 300,000 km every second or 3 x 10^8 ms^{-1}. This means that $v^2/c^2 = 0.000000000000444$ and if you try subtracting this from 1 an ordinary calculator will just give you an answer of 1 because the number is just too small and the calculator is not accurate enough.

If you know something about indices you will know that the formula

$$y = \frac{1}{\sqrt{1 - v^2/c^2}}$$

can also be written as

$$y = (1 - v^2/c^2)^{-1/2}$$

Now there is an exceedingly useful little theorem called the Binomial Theorem which says that when x is much less than one

$$(1 + x)^n = (1 + nx)$$

If we apply this theorem to our formula, we arrive at a much simpler expression for γ (but remember this is only true when v is a lot smaller than c)

$$\gamma = 1 + \tfrac{1}{2}v^2/c^2$$

If you try calculating γ using this formula with a calculator you will still run into difficulty when you add the 1 but you should still be able to confirm that for v = 200 ms-1

$$\gamma = 1.00000000000022$$

What this means is that for every second as measured by the moving clock takes 1.00000000000022 s as measured by the stationary clock. In 10 hours, the difference amounts to

$$10 \times 60 \times 60 \times 0.00000000000022\text{s}$$

which equals

$$0.000000008 \text{ s} \ \text{ or } \ 8 \text{ ns}$$

easily within the capabilities of an atomic clock.

Note that, because the moving clock runs slower than the stationary clock, when the two clocks are compared after the trip, the moving clock will read *less* than the stationary clock by a factor of γ.

'Does this mean that if I travel to a distant star and back, everybody will be older than me when I return?'

You bet it does. Suppose in some future century you choose to visit Alpha Centauri, 4 light years away, travelling in a Thomson Astro-cruiser at 80% of the speed of light. To your twin brother left behind on Earth the journey will take (4 / 0.8 =) 5 years out and 5 years back - ie 10 years in all. From his point of view though (and his point of view is more special than yours because it is he who remains 'stationary' all the time) your clocks run 1.67 times more slowly, so you age by only 10/1.67 = 6 years.

'So how long does the journey actually take? 6 years or 10?'

Both. It takes 6 of your years and 10 of his! You can't really say which time it actually takes. Both points of view are equally valid. On the other hand, every pair of events has what is known as a proper time interval between them, this being the time as measured by a clock which travels between the two events at a constant speed (or as in the case of two events which occur at the same place, is stationary).

The journey under consideration involves three events:

Event A : departure from Earth

Event B : turn-around at Alpha Centauri

Event C : return to Earth

The proper time interval between A and B is the time as measured by your clock ie 3 years. Likewise the proper time interval between B and C is also 3 years. But the proper time interval between A and C is not 6 years, it is 10 years - the time as measured by your stay-at-home twin. You can see that proper times do not necessarily add up. You could if you like define journey time to be the amount of time the voyager experiences, which is the sum of all the proper times for each section of the journey. In this case the journey time is 6 years and the proper time is 10 years, but by moving around fast enough you can make the journey time as small as you like. If you went to Alpha Centauri and back at 99% of the speed of light you could do the trip in just over a year (while your twin brother aged about 8 years); at 99.9% of the speed of light the trip would take less than 5 months. Here are some more examples:

speed (% of c)	proper time (years) =distance / speed	journey time (years / days) =proper time / γ
50%	16.0 years	14 years
80%	10.0 years	6 years
90%	8.9 years	4 years
99%	8.1 years	1 year
99.9%	8.0 years	131 days
99.99%	8.0	41 days
99.999%	8.0	13 days

'Wow! Could you really get to Alpha Centauri and back in a fortnight's holiday?'

Well, you could spend two weeks of your time getting there and back, but your boss back on Earth would be hopping mad, since he will have had to wait eight years for you to return.

Also, the formula we have been using assumes you can accelerate a rocket all the way to 99.999% of the speed of light in no time at all. If you did that, everybody inside would be reduced to jelly! But it might be possible to accelerate a rocket at a more moderate acceleration for a long period of time and gradually build up a large enough speed. Let's look at this possibility.

The 1g rocket problem

Suppose we construct a rocket which can accelerate with a continuous acceleration of 1 g (10 ms^{-2}). We set out from Earth, accelerating at 1 g until we are half way to the star we want to visit; turn round and decelerate at the same rate; have a look at the star for a while; accelerate back again for half the journey; turn round and decelerate all the way home. (Life on board a rocket like this would be just like life on Earth, because the acceleration would give the effect of artificial gravity.)

The formula for the time dilation effect in such an accelerated system is this:

$$\frac{aT}{c} = \sinh\left(\frac{aT'}{c}\right) \quad {}^{1}$$

where T' is the proper time of the journey (ie the time as experienced by the space traveller) and T is the (longer) time experienced by those who stay at home; a is the acceleration of the ship and c is of course the velocity of light.

If we work in units of years and light years, the velocity of light c is, of course, 1 light-year per year.

Now by an extraordinary coincidence, the acceleration due to gravity at the Earth's surface g (10 ms^{-2}) works out to be almost exactly 1 light-year per year^2. Check out the working in the box below. [2]

> 1 year is 365 x 24 x 60 x 60 = 3.15 x 10^7 s
>
> Light travels at 3.00 x 10^8 ms^{-1} therefore 1 light-year (ly) = 9.46 x 10^{15} m
>
> The speed of light is, of course, 1 light-year per year (ly y^{-1})
>
> The acceleration due to gravity at the Earth's surface is 9.8 ms^{-2}.
>
> This is equal to 9.8 x (3.15 x 10^7)^2 / 9.46 x 10^{15} = 1.03 light years per square year (ly y^{-2})

So putting $a = g = 1$ ly y^{-2} and $c = 1$ ly y^{-1} our formula therefore reduces to just

$$T = \sinh(T')$$

where T (and T') is in years.

(For the proof of this and other interesting formulae connected with 1 g accelerated rockets see the Appendix A at the end of the book.)

When working out the time dilation effect we must remember that we must do the calculation for each *quarter* of the journey separately.

Here is a table which tells you how many years will pass on Earth during voyages of different lengths. Owing to the exponential nature of the **sinh** function, the figures rocket up dramatically and you can see that in principle it would be possible for a

---

[1] the **sinh** function - pronounced "shine" - is the hyperbolic sin and is a standard function that behaves in many ways rather like the more familiar sin function. It can be found on many scientific calculators if you press **hyp** before pressing **sin**.

[2] The fact that this is anywhere close to unity is complete coincidence, depending as it does on the completely arbitrary relationship between the strength of the Earth's gravitational field and the length of time it takes for the Earth to go round the Sun. To put it another way, it means is that if a stone could go on falling for ever on Earth, it would reach the speed of light in the same time that it takes the Earth to go round the Sun once – ignoring Relativity, of course!

human being living now to return to Earth within 60 years to see what it will be like in 6½ million years time!

total time on ship (years)	¼ time on ship (years)	¼ time at home (years)	total time at home (years)
6	1.5	2.4	9.4
10	2.5	6	24
20	4	74	297
40	10	11,000	44.000
60	15	1,600,000	6500000

'That's incredible! Could you really travel into the future just by building a fast enough rocket?'

Unfortunately, two practical difficulties stand in the way of this dream. Firstly it will be necessary to build a rocket motor that can sustain an acceleration of 1g for many decades. Sadly, no known or even theoretically possible propulsion system comes anywhere near this requirement. Secondly, a spacecraft travelling at nearly the velocity of light would probably be destroyed by all the microscopic interstellar dust particles slamming into it at nearly the speed of light. As we shall see later, the kinetic energy of a particle the size of a grain of sand travelling at a speed at which $\gamma = 1,000,000$ is equal to that of a 6 megaton bomb!

'I see what you mean. In any case – I still don't really believe it'

Well, you are not alone alone. For many years in the 1950's, respected scientists were still discussing the famous Twins Paradox and even now the Internet is littered with postings purporting to show that the effect is neither possible nor logical. Believe me, though. It is.[3]

Prologue to chapter 2 - At the top of the second hill

The roller coaster has now nearly reached the summit of the second hill and as it slows down momentarily you get a chance to have a quick look round. Over on the right the hands of the big clock appear to be moving normally again but somehow, in the minute it has taken you to negotiate the first big dip, the clock, which appeared to you to run slow when we started the descent has inexplicably skipped a couple of minutes ahead.

Looking down you can see one of the children playing shove-halfpenny give his coin a tremendous whack. All the other children's coins were too big to go down the

---

[3] For a more detailed discussion of the Twins Paradox see my article in Physics Education, **32**, No 5 (September 1997), pp 308-313.

crack but there is something different about this one. It doesn't seem round, it seems oval and, incredibly, it drops right through the crack!

'*How did that happen?*' I hear you cry. The answer is that:

Bizarre consequence number 2

Moving objects shrink along their direction of motion.

Hold on to that Principle! I shout. You are going to need it again! The roller coaster tips violently forward and we are plunging downwards again...

Chapter 2 - Length contraction

Cast your mind back to the river race. Albert takes longer than usual to cross the river because he has to use part of his velocity aiming upstream. Beatrice, however, takes a lot longer because it is really hard work rowing directly upstream.

Now consider the Michelson-Morley experiment. In fact, suppose you are watching someone else performing the experiment in a fast spaceship travelling at nearly the speed of light. The photon which travels at right angles to the direction of motion is a photon clock and, like Albert, it is slowed down by a factor of

$$y = \frac{1}{\sqrt{1 - v^2/c^2}}$$

The photon which moves parallel to the motion of the ship is like Beatrice and ought to be slowed down even more. But as we know, the Michelson-Morley experiment gives a *null* result - ie the two photons take exactly the same time for the journey. (Incidentally there can be no disagreement between you and the people on the space ship as to whether the photons arrive at the same time or not. It would be easy to rig up a device to blow up the ship if any difference was detected and it is not possible for you to disagree with your colleagues on the ship as to whether they are blown up or not. At least, not according to the Theory of Relativity. Now Quantum Theory... well that's another story!)

So how do we reconcile these two viewpoints? The answer is that the photon which moves parallel to the direction of motion of the ship does not have to go as far because lengths *in that direction* are contracted. Here is the formula (for the proof see Appendix B):

$$l = \frac{l_0}{\gamma} = \sqrt{1 - v^2/c^2} \cdot l_0$$

We see that while lengths perpendicular to the direction of motion remain unchanged, lengths parallel to the direction of motion must *decreased* by a factor of γ to preserve the constancy of the velocity of light.

Note that the length contraction factor is just the same as the time dilation factor but you have to divide rather than multiply. For example, if you travel at 80% of the velocity of light past someone at rest, your spaceship will appear to him as if it is only 60% of its *proper* length. Of course, *he* will claim that it is *your* spaceship that is squashed but that's relativity for you!

'*Is that why the penny went down the crack?*' you ask.

Absolutely. The penny was going so fast that it became shorter than the crack and so fell down – see the diagram below:

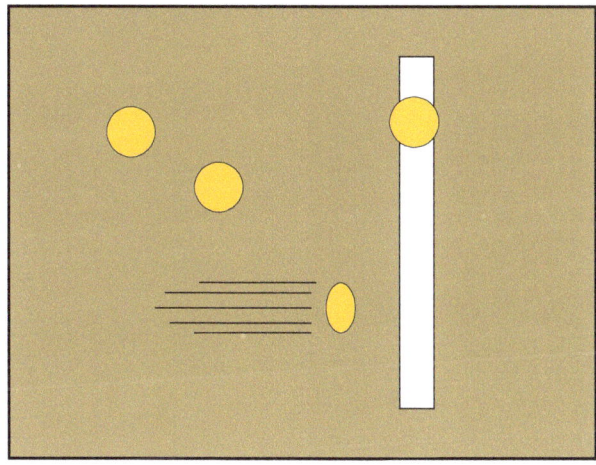

'*Hey - wait a minute! Suppose I was a flea sitting on the penny! Wouldn't it be the crack that was shrunk and not me? I am prepared to accept that times and lengths might* appear *to be extended and shrunk but surely, either the penny goes down the crack or it doesn't! Ha! Ha! Got you now! I knew there was something wrong with this Relativity business!*'

When you have quite finished crowing I will explain.

You mean to say you have an answer to that one as well?'

Uh-uh.

'Well, what is it then?'

I am afraid you will find this even harder to believe than anything I have told you so far.

'Go on - try me'

Well it is like this. From your point of view, as the penny begins to extend over the (contracted) crack, the front of the penny begins to accelerate downwards *before* the back end leaves the table. Like this:

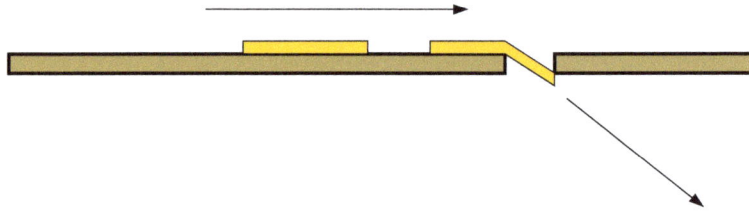

'But the penny can't bend*!'*

Well it doesn't bend in a physical sense, no more than it contracts in a physical sense. But to the flea sitting on the penny, that's what *appears* to happen. Correction. To the flea sitting on the penny, that's what *really* happens. I am afraid we have to accept:

Bizarre Consequence Number 3

Moving objects whose speed changes can appear to bend

'Well you were right after all.'

So you do believe it then?

*'No. I **don't** believe it. But you **were** right when you said I wouldn't believe it!'*

But you **accept** it?

'Yes. I suppose so! But I am getting so confused. Is there anything *that two observers can actually agree on?'*

Yes there is. In fact this is a rather important theorem which I shall call

Reassuringly Normal Consequence Number 4

*Although two observers in relative motion can each argue that they are stationary and that it is the other who is moving, they will **both agree on their relative velocity***

We shall look at this in more detail in the next chapter.

Prologue to chapter 3 - The long straight

After two hair-raising descents, the roller coaster levels out and enters a straight section of track.

Beside the track is a spare train, identical to ours except that, of course, it looks shorter. As I get out my stop watch to time its passing I notice my brother sitting in the other train with watch in hand timing me ...

Chapter 3 - Relative velocity

Suppose my brother and I have two identical twin sister ships, each 100 m long when at rest. If we pass each other with a relative velocity of 80% of the velocity of light, to me my brother's ship appears to be only 60 m long. (g = 1.67 so l '= 100/1.67 = 60 m.)

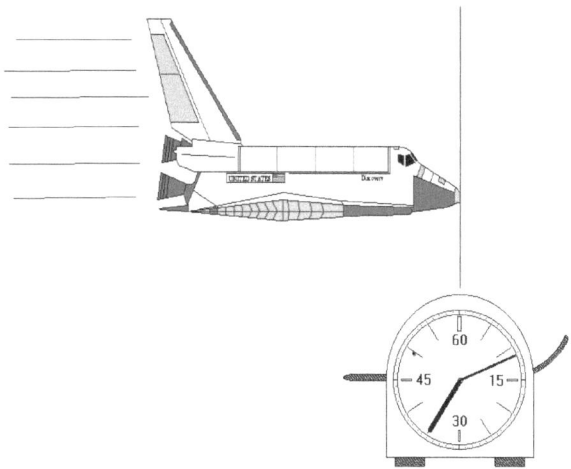

If I time his passing with an accurate clock I will find that he takes 0.25 μs to go by. Using the formula speed = distance / time I calculate his speed to be 60/(0.25 x 10⁻⁶) = 240,000,000 ms⁻¹.

Of course, if my brother makes the same measurements on my ship he will come to exactly the same conclusion about my speed.

But what does *he* think of *my* measurements and *I* of *his*? Lets concentrate on the latter. I watch as he approaches my ship and observe him start his clock at the instant he passes the front of my ship and stop his clock at the instant he passes the back. His clock stops at 0.25 μs. So far so good.

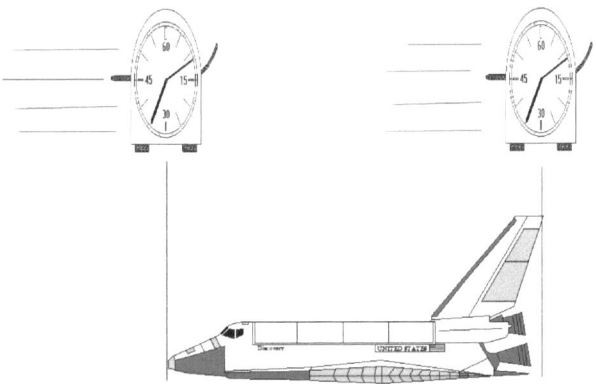

Now I observe him doing his calculations. He starts to press the 6 and the 0 buttons. No that's' wrong I exclaim. *My* ship is not 60 m long it is 100 m long! Then I watch him divide by the time 0.25. No that's wrong too, I exclaim. *Your* clocks are running *slow*! You should be dividing by 0.417 not 0.25! (0.417 is of course 0.25 x 1.67) Grabbing a calculator I perform the calculation for him - only to get... $100/(0.417 \times 10^{-6}) = 240,000,000$ ms^{-1}!

So it doesn't matter who does the calculations, we still get the same result. Each of us is convinced that the other has used the wrong data but we both agree about the answer!

The point is that what is time dilation to me is length contraction to my brother.

Prologue to chapter 4 - The Big Loop

As I glance ahead I see that the roller coaster is hurtling towards one of those up-and-over loops.

'But there is still something puzzling me about the penny.' you say. *'How can **both** ends of the penny start to fall at the same time in one frame of reference while the flea sees the front start to fall before the back? I don't get it.'*

Over on the left, you hear the sound of a train. This one is really shifting and you watch it plunge into the tunnel. The train you saw before had just the same number of carriages and was the same length as the tunnel but you are not surprised now to see that this one seems a lot shorter and there is no sign of the engine emerging when the last carriage plunges out of sight.

Suddenly you hear a couple of loud explosions. Two large clouds of dust rise from the two ends of the tunnel on the railway line. Someone has blown up the tunnel!! A moment later, the engine of the train bursts through the heap of rubble at the exit of the tunnel and you watch horrified as the coach after coach piles into the wreckage. But just at that moment your world begins to turn upside-down…

Chapter 4 - Simultaneity

There is a curious paradox connected with length contraction which I shall call the train in the tunnel paradox. A train whose length when at rest is agreed by all observers to be 100 m enters a tunnel whose length is also exactly 100 m. The train stops in the tunnel and everybody agrees that when the engine is flush with the exit, the guard's van is flush with the entrance.

Some time later, the train passes through the same tunnel at high speed - 80% of the velocity of light in fact. At this speed the g factor is 1.67 and to the man standing by the track at the entrance to the tunnel the train appears to be only 60 m long and spends an appreciable amount of time completely inside the tunnel.

The engine driver, however, sees things rather differently. To him it is the tunnel which is whizzing past at 80% of the velocity of light and which is in consequence shrunk to 60 m in length.

Surely they can't both be right? To put it even more forcibly, suppose the man beside the track is in fact a terrorist and his mission is to trap the train in the tunnel by blowing up the exit at the instant the back of the train enters the entrance? Will he succeed or won't he? To the engine driver, at the instant the back of the train enters the tunnel, the engine is already outside it!

Basically, the whole notion of simultaneity has to be abandoned and we must accept the brutal fact that:

Bizarre Consequence Number 5

Events which occur at different places but at the same time to one observer may happen at different times according to another.

How does this solve the train in the tunnel paradox? We have to ask ourselves how the terrorist is going to arrange his devious trap. He could arrange for an optical signal, triggered by the passage of the last carriage into the tunnel, to blow up the other end of the tunnel - but light does not travel instantaneously and by the time the signal has triggered the explosion, at least part of the train might already have emerged from the tunnel.

What he must do is send the signal *in advance*. We must suppose that he knows how fast the train is going and how short it is going to be, so it is easy for him to *calculate* the time delay needed between the instant the *front* of the train enters the tunnel and the time the signal needs to be sent.

Sure enough, when the time comes, the signal is sent and at the instant the back of the train enters the tunnel, the explosion goes off and the whole train is wrecked.

What does the train driver think of all this? He reasons (quite correctly) that at the instant that (to him) the back of the train enters the tunnel he will be well out of the tunnel and in no danger of being trapped inside by the simultaneous explosions but as he approaches the tunnel, he is horrified to see the terrorist (whose clock appear to be going so *s-l-o-w-l-y*) set off the light signal far too *early*. What is more, the tunnel is so *short* the light signal takes no time at all to get to the far end and the explosion goes off just before the engine reaches the end of the tunnel. Meanwhile, the back of the train (which to the engine driver appears not to have yet entered the tunnel) continues to move as if unaware of the carnage ahead and at the instant the last carriage enters the tunnel the second explosion goes off and the terrorists triumph is complete!

The truth is that events which are simultaneous to the terrorist are *not* simultaneous to the engine driver.

We had better take a careful look at what we mean by events being simultaneous to one observer first of all. When the terrorist says that to him the two explosions were simultaneous, he doesn't mean that he *saw* them happen at the same time. Since he is standing at the tunnel entrance, he sees the explosion happening at the other end (event X) *after* it actually happens. In fact he actually *sees* three distinct events from where he is standing:

Event A: light signal sent to the front of the tunnel

Event B: explosion set off at the back of the tunnel

Event C: light returns from the explosion at the front of the tunnel.

Here is a diagram showing the terrorists point of view:

The terrorist's view of events

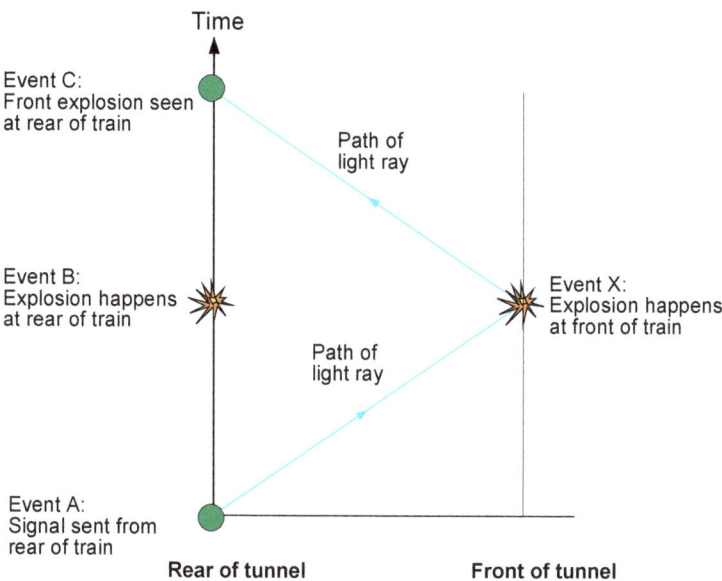

Now, according to the terrorist, event B happens exactly half way between events A and C. It is this equality which justifies his assertion that event X (the explosion at the front of the tunnel) happens at the same time as event B, the explosion at the back.

Why doesn't the driver agree with this analysis? Well, according to him, he is stationary and it is the tunnel which is moving. He sees the light pulse travelling at the speed of light towards the front of the tunnel but he also sees the front of the tunnel rushing towards the light pulse. To him, therefore, the time taken for the light pulse to get to the front of the tunnel is quite short. (This is like Beatrice on her home run.) The light takes an age to return, however, as the tunnel is now travelling in the same direction as the light beam. Of course, the driver will agree that the second explosion (the one at the back) goes off exactly half way between the sending of the signal and the arrival of the light from the first explosion but he will not agree that it is simultaneous with the first explosion. On the contrary, he will maintain that it happens much later.

To the engine driver, it seems as if space and time are no longer at right angles and the 'lines of simultaneity' are tilted forward at an angle. The signal seems to reach the front of the train quite quickly and therefore to him, the explosion at the front seems to occur *before* the explosion at the back.

The driver's view of events

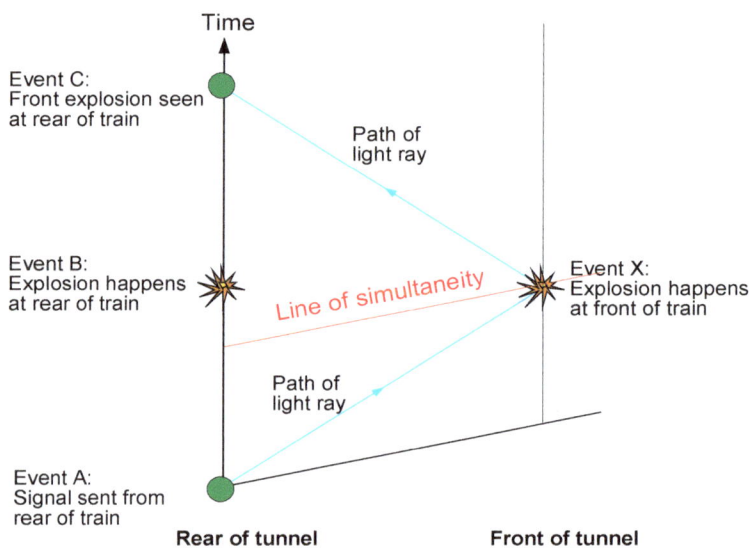

To put it another way, since the engine driver is convinced that the explosion at the front occurs *before* the explosion at the back, the explosion at the front is already history when the explosion at the back happens. So events which happen at places *in front* of you (ie in the direction towards which you are moving) are shifted further into the *past* while events *behind* you are shifted towards the *future*.

Lets pursue this idea a bit further. Two observers who are *stationary* with respect to one another can agree about simultaneous events even if they are a long way apart. It makes perfect sense therefore to think about what is happening *now* on some distant star providing you are not moving with respect to that star. Suppose you have an alien friend who lives on the planet Mishtar in the Rigel star system. Rigel (the brightest star in the constellation Orion) is 1000 light- years away from Earth and, we shall suppose, the planet Mishtar is very like Earth and has a year exactly equal to one Earth year. One day you receive a message saying that your alien friend is this minute celebrating his 5,000 birthday (you must suppose that Mishtarians have discovered the secret of eternal life too!), It makes perfect sense to reason that since the signal has taken 1000 years to reach you, your friend is actually 6,000 years old *now*.

But as you raise your glass in salutation, you suddenly realise that in the month of September, Earth is rushing towards Orion at a speed of 30 km/s (that is 0.0001 light years per year and that therefore you must apply a relativistic correction to this calculation. When you move towards a distant star, your perception of what is happening *now* on that star changes. As we have seen, it is shifted by a certain amount into the *past*. The formula for the *change* in time on that distant star as you move towards it at a velocity *v* is this:

$$\Delta T = \frac{lv}{c^2}$$

where *l* is the distance between you and the star as measured when you are at rest with respect to the star. (for a proof of this formula see Appendix C)

Now, working in years and light years, $v = 0.0001$ light-years per year, and of course $c = 1$ light-year per year, so $\Delta T = 1000 \times 0.0001 = 0.1$ years $= 36.5$ days. So just by moving towards him you can make your Mistarian friend older by 36.5 days! You are too late! You should have celebrated his birthday a month ago!

The time change may not sound much but the effect increases with distance so time on a galaxy 1,000,000 light years away varies by as much as 100 years as the Earth swings around its orbit about the Sun. And even *walking* towards a galaxy 1 billion light years away changes time there by 5 years!

Of course, like time dilation and length contraction, the effect is not real. You can't actually make someone older by moving towards them. What we are saying is that when you change you speed, you also completely change your perspective of what is simultaneous.

You can explain the Twin's Paradox like this. From the point of view of your twin brother on stationary Earth, you (on your Thomson Astro-cruise to Alpha Centauri at 80% of the velocity of light) simply age more slowly than he does and return 10 years later only 6 years older. From your point of view, it is *he* who ages more slowly and in the first 3 outward years of your journey, *he* will only age by 1.8 years. But when you turn round, *your* perception of what is *now* suddenly changes by DT = lv/c^2 = 4 x (2 x 0.8) = 6.4 years. (The factor of 2 is because your speed changes from -0.8c to +0.8c.) When you get back, you find that your brother is $1.8 + 6.4 + 1.8 = 10$ years older!

The question of what is *now* on a distant star is, of course, quite academic because the whole point is that you are not there to check it! but it might occur to you that if you can change the *time* at which events on a distant planet seem to occur just my moving around, you might be able to change the *order in which events at different places occur*. Well, as a matter of fact you can. If your friend on the distant star (Λ) happens to have a (slightly older) twin brother who lives on another star (B) 1000 light years away in the opposite direction. By moving away from star B and towards star A you can make the younger brother older than the elder brother! Or to put it another way, you can arrange it so that (to you) the younger brothers birthday happens *before* the elder brothers birthday!

'Wait a minute' you say. *'If you can change the order in which events occur, is it possible to make the **effect** precede its **cause**? We have met four **bizarre***

consequences so far, could this be the fifth? If so, you could travel backward in time . . . You could go and kill your own grandmother . . .'

No you couldn't! Relativity may be bizarre but it is not illogical! You can indeed alter the order of events but only if they happen *in different places*. What is more, the events have to be fairly close together in time as well. To be more specific, you can only reverse the order of two events if a light beam triggered by event A cannot reach B before event B happens. We can summarise this as follows:

Reassuringly Normal Consequence Number 6

*Although the motion of an observer can alter the order in which some events appear to happen, **relativity is entirely consistent with the laws of cause and effect**.*

Mind you, this does imply one terribly important consequence which we must add to our list:

Interesting Consequence Number 7

It is absolutely impossible to send a signal of any sort at a speed faster than that of light in a perfect vacuum.

'Why? What would happen if you could?'

Well, let us suppose that at some time in the future, your friend on Mishtar tells you the secret of instantaneous telepathy. One day you get a telepathic message from Mishtar with wonderful news. Your friend has reached the grand age of 10,000 years and he is giving a big party to all his friends *right now*. Of course, you respond by instantly conveying your congratulations. Nothing wrong in that *provided that you and your friend are stationary* with respect to each other.

But suppose that, having received the message, you get into a spaceship and accelerate in the direction *away* from Orion. If moving *towards* a distant star makes your friend *older*, moving *away* from it will make him *younger* (always from your point of view, of course). If you were to accelerate to a speed of 30 km s^{-1}, you would make him 36.5 days younger and if you sent your congratulations to him now, he would receive the reply before he had sent his initial message! If this sort of thing was possible, then, with the aid of an accomplice on a distant planet, you could effectively predict the future by telling your distant friend what has just happened and getting him to return the information several days earlier! This is nonsense!

The fact is - no *signal* of any sort can travel faster than the speed of light.

If we have to abandon the notion of simultaneity, we also have to rethink the concepts of past, present and future. Our usual picture of the nature of time can be represented something like this:

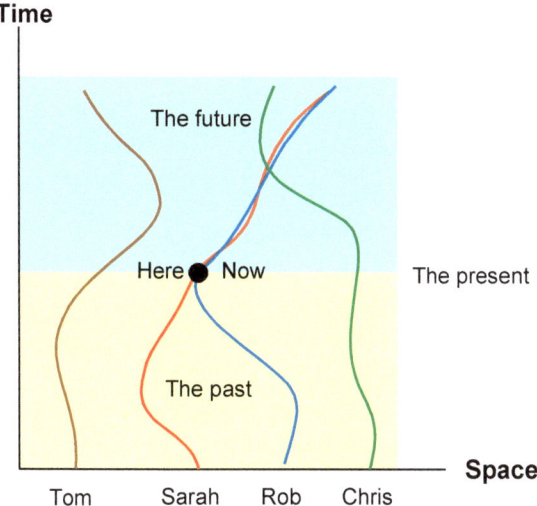

The 'present' is a line which moves inexorably up the diagram gradually turning 'future' events in to 'past' events. Four characters in this simple universe are represented by 'world lines' which snake their way up the diagram. For Sarah and Rob, the point labelled 'Here/Now' appears to have special significance. (Note that although I have extended the world lines into the future, this is not intended to imply that the future is in any way pre-determined. That is another question entirely.)

Relativity forces us to take a rather different view. Since different observers have a different perception of now, we must divide spacetime into four regions. the past, the future and what I shall call the inaccessible past and the inaccessible future. Past and future are therefore contained within an hourglass shape whose limits represent the speed of light (On the diagram below, the speed of light is represented by a 45^0 line.)

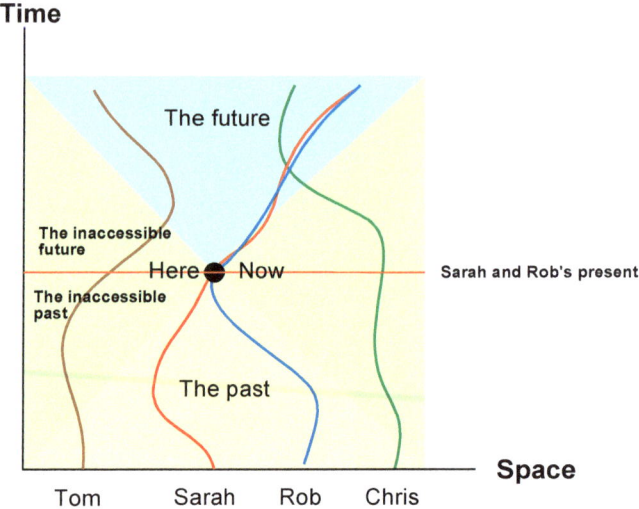

For any individual at a point in spacetime (called the here/now) the past contains all events which could in principle have a causal effect on the here/now by a signal which travels no faster than light. The inaccessible past contains all those events which happened so recently and/or are so far away that no signal could possibly reach the here/now from them. Likewise the future is the collection of all events which could, in principle be causally affected by a signal sent from the here/now while the inaccessible future is the collection of events which have not happened yet but which we cannot in any way influence because light does not have time to reach them.

If we consider Sarah and Rob to be at rest when they meet, their now is a horizontal line. But if they move say to the right, relativity causes their now to incline up wards like this:

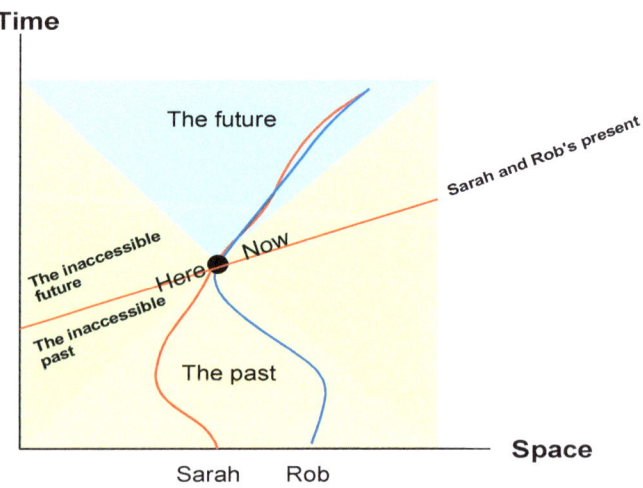

In fact it works rather like a pair of scissors; If the world line makes an angle of 20^0 with the vertical time axis, the now line makes an angle of 20^0 to the horizontal space axis. Note that by moving with the right speed, any point in the inaccessible past or inaccessible future can be made simultaneous with here/now. In a sense then, both the inaccessible past and the inaccessible future must be considered to be part of the potential present.

Note that at the instant Sarah and Rob meet (which happens to be on a spaceship moving to the right), Tom and Chris can be anywhere in this potential present. As it happens, both Tom and Chris are stationary (ie their world lines are vertical) where their world lines intersect Sarah and Rob's now.

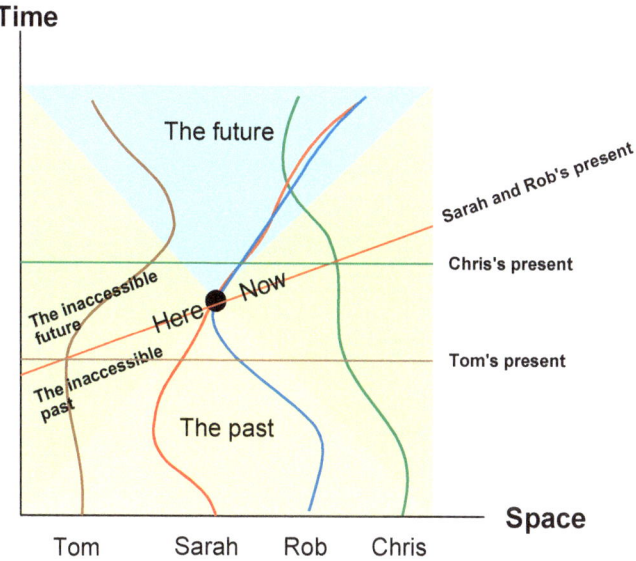

If Sarah could have an instantaneous telepathic conversation with Tom it might go something like this:

"Hi Tom! I have just met this great guy called Rob at a New Year's party last night."

"Well that's great news, Sarah, and I sure hope you get along - but surely New Year's day on Earth is next week?"

Likewise, if Rob sends a telepathic message to Chris, she may be delighted at the news but will consider it as history.

Of course, the vital thing to appreciate is that faster than light messages are forbidden. If all messages travel at the speed of light or less, there is no illogicality because all events are eventually seen to happen *in the same order* from all observers points of view. We do, however, have to accept that there is no such thing as 'the present'. Each observer has a now, but different observers simply have different nows.

'I see. It is as if past and future have some fuzzyness where they meet'

That's right. But they always come together in sharp distinction at the here/now. Some authors, (eg Rudy Rucker in 'The Fourth Dimension' Penguin, 1985) have taken the blurring of past present and future to imply that the passage of time is an illusion. In fact this is an age old idea and I do not believe that Special Relativity lends any weight to it at all. Each of us has a past and each of us has a future. The fact that we cannot always agree on what exactly constitutes the present is just another of those bizarre consequences which we just have to accept.

Prologue to chapter 5 - The Switchback

It is quite a relief after toying with the idea that effects might precede their cause to see the world right itself and for things to look normal again as the roller coaster enters a high speed switchback. From this vantage point you can see that the boys on the rifle range are playing some funny kind of game. They are running towards the target as fast as they can while firing their rifles . . .

'Seeing those boys has given me an idea.' you say. *'Suppose I have a gun which can fire bullets at 80% of the speed of light and suppose I fire them from the front of my spaceship which is travelling at 80% of the speed of light. Surely the bullets will be travelling faster than light then?'*

Good idea, but speeds don't add up like that in relativity. In fact there is an important theorem about the addition of velocities which I shall call:

Bizarre Consequence Number 8

Speeds do not add up in the usual way and the result of adding two speeds together is always less than the speed of light however great the original speeds are.

Chapter 5 - Adding speeds together

If *I* see your spaceship travelling past at a speed *u* and *you* see your bullets travel away from your ship at a speed *v* it turns out that *I* will see *your* bullets travelling at a speed:

$$w \; = \; \frac{u+v}{1+uv/c^2}$$

(For a proof of this formula see Appendix D)

If either u or v is much smaller than the speed of light the term $1 + uv/c^2$ on the bottom is essentially equal to 1 and the formula reduces to $w = u + v$ which is what we would expect.

Plotted as a three dimensional graph, you can see that the speed of light acts as a kind of ceiling above which it is impossible to go.

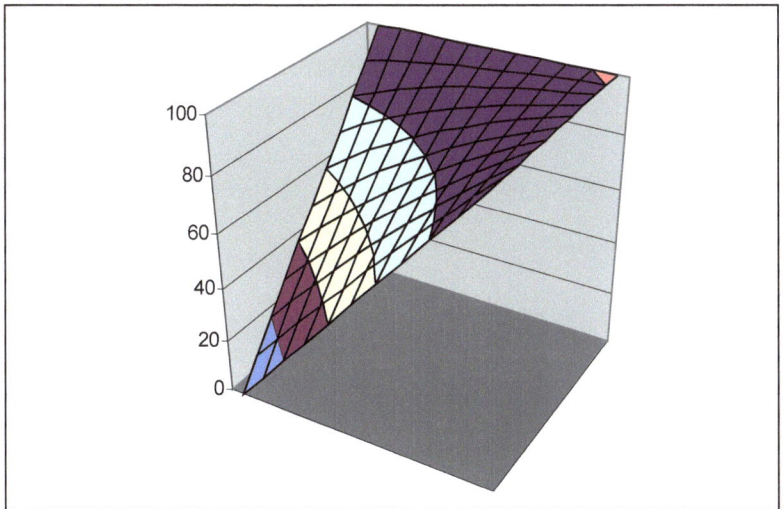

Suppose you had a multistage rocket each of whose stages could reach 80% of the speed of light. A two stage rocket would reach 80% ++ 80% = 97.56%. A three stage rocket would reach 97.56% ++ 80% = 99.72% and a four stage rocket, 99.97% of the speed of light. In fact you would need an infinite number of stages to reach 100%!

'But what happens to the 1g rocket? That goes on accelerating at a constant rate so it must be getting faster and faster. Surely then it must eventually reach and exceed the speed of light, mustn't it?'

You're right about it getting faster and faster with respect to the universe outside and it is also true to say that from the point of view of the occupants of the rocket, the acceleration is constant in the sense that they experience a constant artificial gravity of 1g. But the rocket never exceeds the speed of light. It turns out that the speed of the rocket increases according to the formula

$$v \; = \; c \tanh \frac{a}{c} t$$

(The details are given in Appendix A on the 1g rocket problem.)

This formula describes a curve which increases exponentially to a limit and for a rocket accelerating at 1g, it looks like this:

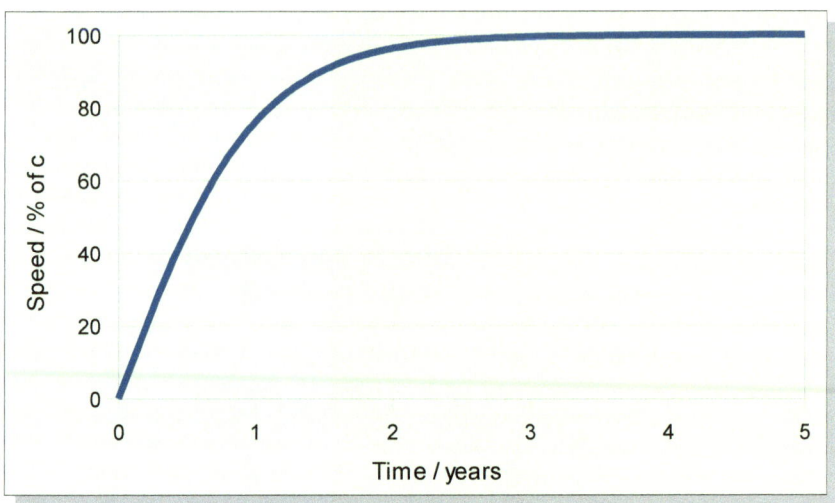

Just in case you think you are getting the hang of this relativity business, I should like to remind you of another consequence of Special Relativity.

Bizarre Consequence Number 9

Although it is impossible to travel faster than the speed of light, it is perfectly possible to travel to a distant star in less time than it takes for light to get there.

'Oh, Come on!' you cry, *'That doesn't make sense at all! You have just proved to me that you can't travel faster than light so how can you say, in the next sentence, that you can travel faster than light!'*

But I didn't say that.

'Yes you did'

No, I said it was perfectly possible to travel to a distant star *in less time than it takes light to get there.*

'Well that's just the same thing!'

No it isn't because it depends on *whose time* we are talking about. I will admit that I did phrase my statement in such a way as to make it sound a bit paradoxical but the truth of the matter is that if you travel to Alpha Centauri at 80% of the speed of light, you will get there in 3 years - 3 of *your* years, that is. Light takes 4 years to get there. (For more details, see Appendix E)

'Well, aren't you travelling faster than light then?'

No. If you set out at the same time as a beam of light, the light will still get there first because it is travelling faster than you. That it only takes 3 of your years to get there can be explained in either of two ways. From *my* point of view (left behind on Earth) your clocks are ticking slowly because of time dilation. From *your* point of view the distance from Earth to Alpha Centauri is shrunk because of length contraction. Which explanation you chose depends on your point of view.

*'So how long does it take light to get there **from the light beams point of view?**'*

Now there you do have an interesting question. In fact it was this very question that started Einstein thinking about the whole business when he was still a student. At the speed of light, time stops still and the whole universe shrinks to a point. I suppose it is meaningless to contemplate what the world looks like from a photon's point of view. Indeed, I rather doubt if a photon *has* a point of view; but if we must pursue the idea at all we have to conclude that, to a photon, the whole universe is nothing but a single point which exists for zero time. If God exists, He is a photon.

Prologue to chapter 6 - The spiral

The roller coaster now enters a series of tight curves which cause it to go round in a horizontal circle faster and faster and faster. Some distance away a siren goes off and you are not surprised to hear the pitch of the siren wail up and down as first you approach and then recede from it.

'Well, at least the Doppler effect works as normal' you say.

Actually, it doesn't work quite as normal. There are, in fact two normal Doppler effects, the 'moving source' effect and the 'moving observer' effect. But in light there is only one Doppler effect and it isn't quite the same as either of them!

> ## Interesting Consequence Number 10
>
> *There is only one Doppler effect in light.*

Chapter 6 - The Doppler shift

When you hear an ambulance go by, the pitch of its siren falls. Since the ambulance is moving through the air and it is you who is stationary, this is due to the moving source effect. It works like this.

Suppose that the ambulance s moving away from you and its siren has a period T_0. During the time it takes for the siren to emit one complete wave, the wave front has moved towards you a distance cT_0 but the ambulance has moved in the opposite direction a distance vT_0. The wave is therefore *stretched* to a total distance $(cT_0 + vT_0)$

Since this wave reaches me travelling at a speed of c, the time T it takes for the wave to pass me is $(cT_0 + vT_0)/c$ *ie:*

$$T = T_0(1+v/c)$$

The apparent time period has been *increased* by a factor of $(1 + v/c)$.

(If you are a musician, you will know that a change of one semitone – which is $1/12^{th}$ of an octave – is caused by a change in pitch of $^{12}\sqrt{2}$ or 1.06 This means that an ambulance travelling at 45 mph (which is 6% of the speed of sound) will have its siren Doppler shifted by a whole tone as it goes by – a semitone up as it approaches and a semitone down as it recedes.)

Now exactly the same thing happens with light. Hadn't you noticed that in addition to the siren sounding lower, all the lights on the ambulance look redder too? No? Well I am not too surprised as the ambulance may be travelling at an appreciable fraction of the speed of sound, but it is not travelling anything like as fast as the speed of *light* so the effect is not going to be very noticeable.

The argument in light is exactly the same except that, in addition to the increase due to the wavelength stretching effect, we must also include the increase due to time dilation. All we have to do is replace T_0 by γT_0 ie:

$$T = \gamma T_0(1+v/c) = T_0(1+v/c)/\sqrt{1-v^2/c^2}$$

Multiplying top and bottom by c and cancelling a factor of $\sqrt{(c + v)}$ leads finally to

$$T = T_0\sqrt{\frac{c+v}{c-v}}$$

'But I thought you said that there was only one Doppler effect in light. You seem to have found a formula for the moving source effect – but what about the moving observer effect?'

OK. Consider the normal effect (in sound) first. Suppose you are blowing a whistle to attract the attention of the ambulance driver (who is still travelling away from you). The waves you send (which have a wavelength cT_0) are chasing the ambulance with a closing speed of $(c - v)$. The time it takes for one wave to catch up the ambulance is going to be $cT_0 / (c – v)$

$$T = \frac{T_0}{1-v/c}$$

Again the time has been *increased* so the ambulance driver will hear a lower pitched whistle, but you will notice that the formula is not the same as the formula for the moving source effect which *multiplies* by $(1 + v/c)$ rather than *dividing* by $(1 - v/c)$.

Now we must be a little careful when applying the relativistic time dilation correction here. From our point of view, it is the ambulance man's clock which is going slow. So if it takes T of *our* seconds to reach the ambulance, it will take *fewer* of *his*. This means that we must *divide* by γ, not multiply. Hence

$$T = \frac{T_0}{(1-v/c)\gamma} = \frac{T_0\sqrt{1-v^2/c^2}}{(1-v/c)}$$

which, surprise, surprise, leads us to exactly the same formula!

Astronomers are usually more interested in changes in wavelength rather than changes in time period, but the formula is essentially the same as wavelength is proportional to period. ie:

$$\lambda = \lambda_0 \sqrt{\frac{c+v}{c-v}}$$

I find it rather pleasing that the formula for the Doppler shift in light turns out to be the geometric mean of the two normal Doppler shift formulae. Plotting some graphs will help to sort out the differences.

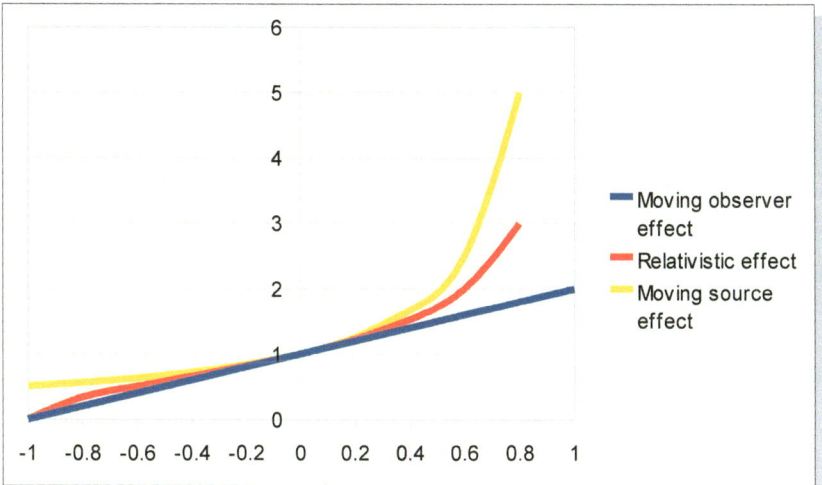

From a physicists point of view, it is the *Doppler shift factor* here defined as λ/λ_0 which is of primary interest. From an astronomer's point of view, it is the *change* in wavelength $\lambda - \lambda_0$ which is generally measured and they usually quote a star's *red-shift factor* defined as $(\lambda - \lambda_0)/\lambda_0$. It is easy to see that *Red-shift factor = Doppler shift factor - 1*

If the speed of the source is much smaller than the speed of light, we can simplify the formula using the binomial theorem as follows:

33

$$\frac{\lambda}{\lambda_0} = \sqrt{\left(\frac{c+v}{c-v}\right)} = \sqrt{\left(\frac{1+v/c}{1-v/c}\right)}$$

$$\frac{\lambda}{\lambda_0} = (1+v/c)^{\frac{1}{2}}(1-v/c)^{-\frac{1}{2}} \approx (1+\tfrac{1}{2}v/c)(1+\tfrac{1}{2}v/c) \approx 1+v/c$$

hence:

$$\text{Red-shift} \approx v/c$$

This expression can only be used in the range where the speed of the galaxy is within 20% of the speed of light. The Andromeda galaxy, for example, shows a red-shift of -0.001 (ie it is in fact *blue*-shifted) and is moving *towards* us at a speed of about 300 km/s. Don't worry about a collision though – at 2.5 million light years away it is going to take 2.5 billion years to reach us!

Quasars with very large red-shift factors have been observed – for example, quasar PC1247+3406 has a red shift of 4.897 therefore a Doppler shift factor of 5.897 and a recession speed of 94% of the speed of light. In 1929, Edwin Hubble discovered that all the distant galaxies were receding from us and subsequent measurements have indicated there is a simple proportional relationship between speed and distance with the edge of the observable universe at about 13.6 billion light years. If PC1247's red shift is due toe the Doppler effect and nothing else, it would seem to indicate that it lies a staggering 12.8 billion light years away.

Of course, if a galaxy were to travel away from us at speed equal or greater than that of light, it would be invisible because all light from it would be red-shifted out of existence. As it happens, it is not inconceivable that a galaxy should recede from us at such a speed in spite of what we have said about the addition of velocities etc. but this can only happen in a evolving universe, not a fixed one.

Prologue to chapter 7 - The third hill

The roller coaster, still travelling at high speed, emerges from the spiral and shoots up a gentle rise. Once again, as it approaches the summit you have a chance to catch your breath and look around. There are those two boys with their two footballs beside the tracks of the roller coaster. What they are doing is really rather clever. Instead of just playing catch, they each toss their ball towards the other with just the right speed and direction so that the balls bounce off each other and return exactly where they came from.

But you don't have time to watch much longer. The roller coaster is starting its third big descent. Hang on!

To your surprise, you discover a football in your lap just like the ones the boys were throwing. On a whim you toss it sideways out of the coaster. Just then, ahead of you, you see one of the boys throwing his ball towards the tracks. To your astonishment, you realize that the two balls are on a collision course but your ball seems to be travelling faster than his.

On the other hand, to the boy beside the tracks, it is you who are moving swiftly by and (because your actions are slowed down by time dilation) it seems to be his ball which is travelling faster than yours. Each of you predicts that the balls will bounce asymmetrically towards the other. What in fact happens is that the balls bounce off each other perfectly and the boy catches his ball while your ball lands in your lap! This all happens because of:

> ### *Bizarre Consequence Number 11*
>
> *Moving objects increase their mass.*

Chapter 7 - Mass Inflation

If two perfectly elastic bodies of equal mass m travel towards each other with equal and opposite speeds u, they will bounce off each other with identical but reversed velocity. This is a straightforward consequence of the laws of conservation of energy and momentum.

Now suppose that one of the bodies is also travelling across the line of the collision at a speed of v. The collision will look something like this:

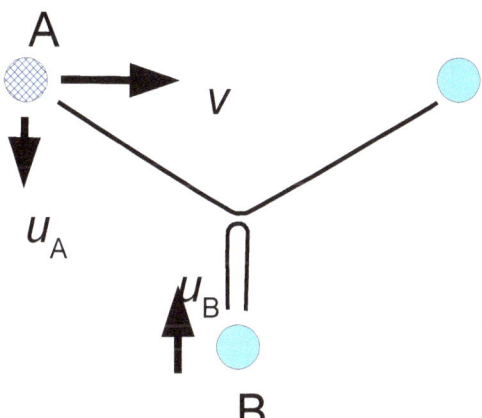

In Newtonian dynamics, the sideways motion of one body is irrelevant and does not change the behaviour of the two balls in the perpendicular direction.

Nor should it in Special Relativity either, even if v is a sizeable fraction of the speed of light. The problem comes when we consider *how* the body is given its sideways velocity u (assumed to be much smaller than v). You see, from B's point of view, the mechanism which propels ball A sideways is running slow because of time dilation and u_A appears to be *less* than u_B by a factor of γ. Naturally from A's point of view it is u_B which appears to be *less* than u_A. All other factors being equal, A and B will therefore predict different outcomes. In fact, since the whole situation must be symmetrical with respect to A and B, we know that what actually happens is that the balls bounce off each other with unchanged speeds.

So what is the solution? Clearly, while the *speed* of the two balls appear to be different, the *momentum* of the balls must be the same. Now momentum is mass x speed and if the speed is apparently reduced by a factor of g because of time dilation, the *mass must be increased by a factor of γ* to compensate. This phenomenon is called mass inflation and the appropriate formula is:

$$M \; = \; \gamma M_0 \; = \; \frac{M_0}{\sqrt{1 - v^2/c^2}}$$

where M_0 is the *rest mass* of the object - ie its mass when measured at rest.

Likewise the *momentum p* of an object moving with a relativistic speed v is equal to:

$$p \; = \; Mv \; = \; \gamma M_0 v$$

This gives us another explanation as to why the speed of the 1g rocket never gets greater than the speed of light. From the point of view of an observer on Earth, the faster the rocket goes, the more massive it gets. Since we assume that the thrust of the rocket is constant, the effective acceleration (as seen from a stationary observer *outside* the craft) must decrease more and more. In fact this is exactly what we observe when we accelerate electrons in a linear accelerator over millions of volts. When they get up to 99% of the speed of light, the extra volts don't make them go much *faster* but they do impart extra *momentum* and of course, *energy* to the electrons.

'Ok, so I accept that mass increases and that the momentum of a particle of rest mass M_0 travelling at a speed v is $\gamma M_0 v$. What about kinetic energy? I suppose that is $\frac{1}{2}\gamma M_0 v^2$. Is that right?'

Well that is an excellent guess. At low speeds, $\gamma = 1$ so the formula reduces to $\frac{1}{2}M_0 v^2$ which is correct and at high speeds γ tends towards infinity so the kinetic energy also becomes infinite which is correct again. Nevertheless, I am afraid it still isn't quite right. The correct answer is:

$$KE_r = M_0 c^2 (\gamma - 1)$$

where Ke_r is the relativistic kinetic energy. (for the proof of this equation see Appendix F)

It is worth writing out both these equations in full and expanding them using the binomial theorem so that the difference is obvious:

$$KE_1 = \tfrac{1}{2}\gamma M_0 v^2$$

$$KE_2 = M_0 c^2 (\gamma - 1)$$

$$KE_1 = \frac{\tfrac{1}{2} M_0 v^2}{\sqrt{1 - v^2/c^2}}$$

$$KE_2 = M_0 c^2 \left(\frac{1}{\sqrt{1 - v^2/c^2}} - 1 \right)$$

$$KE_1 = \tfrac{1}{2} M_0 v^2 \left(1 - \frac{v^2}{c^2} \right)^{-\tfrac{1}{2}}$$

$$KE_2 = M_0 c^2 \left(\left(1 - \frac{v^2}{c^2} \right)^{-\tfrac{1}{2}} - 1 \right)$$

$$KE_1 = \tfrac{1}{2} M_0 v^2 \left(1 + \frac{v^2}{2c^2} + \frac{3v^4}{8c^4} + \frac{5v^6}{16c^6} + \dots \right)$$

$$KE_2 = M_0 c^2 \left(\left(1 + \frac{v^2}{2c^2} + \frac{3v^4}{8c^4} + \frac{5v^6}{16c^6} + \dots \right) - 1 \right)$$

$$KE_1 = \tfrac{1}{2} M_0 \left(v^2 + \frac{v^4}{2c^2} + \frac{3v^6}{8c^4} + \frac{5v^8}{16c^6} + \dots \right)$$

$$KE_2 = \tfrac{1}{2} M_0 \left(v^2 + \frac{3v^4}{4c^2} + \frac{5v^6}{8c^4} + \frac{35v^8}{64c^6} \dots \right)$$

As you can see, the differences are subtle but that is the way it is. Notice that both expressions reduce to $\tfrac{1}{2} M_0 v^2$ for small values of v.)

37

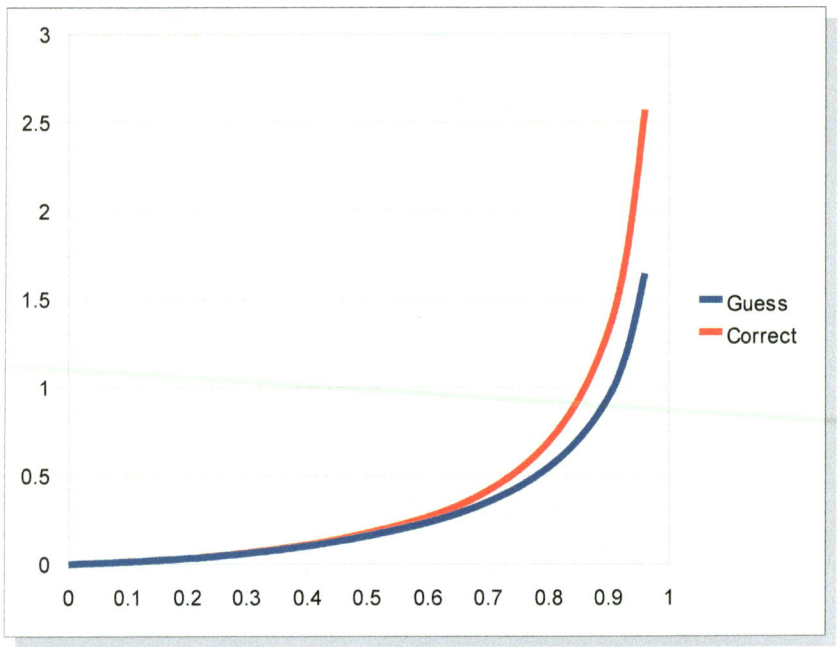

When the two functions are plotted on top of one another you can see just how little difference there is:

The importance of the distinction lies not in the slight differences in numerical values that are generated but in the whole new perspective that the equation throws on the nature of mass and energy.

Let's rewrite the equation in the following way:

$$KE_r = M_0 c^2 (\gamma - 1)$$

$$KE_r = \gamma M_0 c^2 - M_0 c^2$$

$$KE_r = M c^2 - M_0 c^2$$

$$M c^2 = M_0 c^2 + KE_r$$

Now rewriting an equation in a different way doesn't prove anything but it can suggest ideas. Lets give the terms some meaningful names.

Mc^2 is a kind of energy term formed from the *total relativistic mass M*. It seems sensible therefore to call this term the *total relativistic energy* of the moving body *E*.

Similarly, since M_0 is the *rest-mass* of the body, $M_0 c^2$ should rightly be called the *rest- mass energy* of the body E_0.

This permits us to say the following:

Total relativistic energy = rest-mass energy + relativistic kinetic energy

or in symbols:
$$E = E_0 + KE_r$$

where
$$E = \gamma M_0 c^2$$

It was a stroke of genius on Einstein's part to see that the two terms $E = Mc^2$ and $E_0 = M_0c^2$ were not just mathematical junk, they actually had physical meaning. An object at rest really does have an incredible amount of energy locked up inside it. Special Relativity does not prove that a mass at rest has energy since there is no process dealt with by the theory which could possibly release this energy but Einstein speculated on the possibility that one day such a process might be discovered, a speculation which became all too true in his own life time.

So now we have arrived at:

> ### Profound Consequence Number 12
>
> *All massive objects contain energy according to the famous relation:*
>
> $$E = mc^2$$

The significance of this equation cannot be exaggerated. It ranks alongside the discovery of the law of gravity, the idea that zero is a number, the invention of language and the discovery of fire as a turning point in human history - and as an icon for the achievements of the twentieth century, it cannot be surpassed (Hence the flashy border!). Let us observe a few moments of thoughtful and respectful silence before it.

. . .

It hardly needs saying that since the speed of light is quite large, the rest-mass energy of a kilogram of matter is very large indeed. In fact it is equal to 90,000,000,000,000,000 J. The average human being converts energy at a rate of about 100 W so this quantity of energy would keep you alive for about 28 million years! More realistically, it is equal to the total energy output of a large modern power station in 2 years of continuous operation.

On the other hand, the Sun (which outputs a prodigious 4×10^{26} W) eats up 4 *million tonnes* of matter *every second*!

Mind you, that is only a titbit compared to the most energetic objects in the known universe, the **quasars**, which probably emit something like 10^{39} W and eat up the *mass of the Moon every second*!

'*Incredible! So everything that has mass has energy does it?*'

Absolutely right. But that is only really half the story. Einstein's equation works the other way round as well and everything which has energy, also has mass. What I am saying is that a new battery is more massive than an old flat battery, a wound up watch is heavier than a run-down watch, a hot cup of tea weighs more than a cold cup of tea. And things which lose energy get lighter: the total mass of the products of an exothermic chemical reaction is less (after the heat has escaped) than the total mass of the reagents, two magnets stuck together weigh less than the two magnets separately, and most important of all, the mass of an atom is measurably less than the sum of the masses of all its constituent particles.

Even photons, which since they travel at the speed of light have zero rest mass, must have (relativistic) mass by virtue of the energy they contain. Not only that, they have momentum as well, but before we work that out, we have one more important consequence to state which is absolutely true for *all* objects under *all* circumstances. It relates the relativistic energy E of a body to its momentum p and it looks like this:

Important Consequence Number 13

The total relativistic energy E of a body and its relativistic momentum $\mathbf{p}$ are related by the equation:

$$E^2 - p^2 c^2 = M_0^2 c^4$$

(you will find the simple proof in Appendix G)

As we mentioned earlier, this relation has special relevance for the photon (whose rest-mass is zero and for which the usual equations for energy and momentum are undefined). Putting $M_0 = 0$ we get

Important Consequence Number 14

The total energy E of a photon is related its momentum p by the equation:

$$E = pc$$

'What do you mean when you say that a photon has momentum? Surely something which doesn't have mass can't have momentum, can it?'

Well, yes, it can. If you like you can think of it as the momentum of the mass of the energy which the photon carries. To be sure a photon does not have a lot of momentum. If you shine a 1 W laser beam at a black surface, every second the surface absorbs E/c units of momentum ie 3.3×10^{-9} Ns. Since rate of change of momentum equals force, the laser beam exerts a force of 3.3 nN on the surface. Not a lot!

On the other hand, the energy density of solar radiation is 3,000 Wm^{-2} so the force on a solar wind satellite whose sail has dimensions 30 m × 30 m would be 9 mN (9×10^{-3} N). This doesn't sound much either but the force of gravity *from the Sun* on a 1 kg mass at the same distance from the Sun as the Earth is only 6 mN (6×10^{-3} N). So by balancing the force of gravity against the force of the solar wind, it may be possible one day to 'sail' around the solar system for free!

Prologue to chapter 8 - The second roller coaster

At last, the roller coaster jerks to a halt in the station, and you stagger off the train, your mind reeling with important principles and bizarre consequences.

'Well, I survived!' you exclaim *'but oh how my head is spinning!'*

I am afraid we have not quite finished yet though. Gently, I lead you by the arm round a corner where we stop in front of a huge gateway over which is blazoned in scary letters the single word OBLIVION.

'What's this?' you ask.

It is another roller coaster.

'Hell's teeth! Is it as bad as the other one?'

Well, not so bad really - but I have to admit it is a bit of a cheat. There is another roller coaster which is such a rough ride that hardly anyone can stomach it but this one has been specially designed to give you just a flavour of what the real one is like. Are you ready to try it?

'I suppose so. But why is it called OBLIVION?'

Wait and see! But first, while we wait in the queue, I must tell you about Einstein's second Big Idea. The Fundamental Principle of Special Relativity (Einstein's first Big Idea) dealt with observers in uniform motion. The second Fundamental Principle (on which the General Theory of Relativity is based) deals with observers in accelerated motion. It is this.

> ## The Fundamental Principle of General Relativity
>
> *The laws of Physics in a uniformly accelerated frame of reference are identical to the laws of physics in a gravitational field.*

or, to put it another way:

> *It is impossible to carry out any experiment inside a closed laboratory which will determine whether the laboratory is being accelerated uniformly or whether it is situated in a gravitational field.*

'Well that's silly. Surely you can always tell if there is gravity. All you have got to do is drop something. If it falls - there is gravity!'

Not at all. Suppose you are in deep space on a journey to a distant star perhaps. You may be moving, you may be stationary, it doesn't matter. All that matters is that there are no stars or planets nearby so there is no gravity. You have just finished reading a chapter of your favourite book - 'The Hitch-hikers Guide To The Galaxy' - and, as is your custom you simply park it in front of you where, because of the weightless conditions inside the spaceship, it hovers obediently. Suddenly, it falls to the back of the ship, accelerating as it goes. At the same time, you feel your own body pressed into the couch on which you have been lying. What are you to think? Have you suddenly arrived at your destination where gravity is normal again? Possible perhaps. But it is a lot more likely that the captain of the ship has fired the rocket engines and that you and the rocket are simply accelerating. The book did not fall because of gravity. In fact it didn't really fall at all. It simply stayed where it was and the rocket accelerated forwards and caught up with it. Likewise, you weren't pressed into the couch by the force of gravity, no, it was simply the thrust of the rocket engines, transferred to you through the couch which made you accelerate forwards with the rocket.

So the effects of acceleration look, on the face of it, to be just like the effects of gravity. But are they? Is there *any* way to tell the difference? To answer this question, we must have a closer look at exactly what we mean by *mass*.

Has it ever struck you as rather odd that two objects of different mass should fall with the same acceleration in a gravitational field? If not, then you must be either very clever or you haven't thought about it at all. After all, the great Greek scientist and philosopher Aristotle thought that it was self evidently obvious that heavy objects would fall faster than light ones and he carried the weight of intellectual opinion with him for over two thousand years. It was Galileo who first saw the illogicality in Aristotle's theory. His argument went something like this.

"According to Aristotle, a heavy stone should fall twice as fast as a light stone. Now if you tie a heavy stone to a light stone, the heavy stone will pull the light one down faster while the light one will tend to stop the heavy one from travelling so fast. The combination will therefore fall at a speed which is intermediate between the speeds of the two stones on their own. But hang on a minute - if you tie a light stone to a heavy one, surely you make a stone which is heavier than either and therefore, according to the theory, should fall faster than either? There is an inconsistency here, therefore the original premise is false."

Fifty years later, Isaac Newton vindicated Galileo by setting out a wonderful theory of mechanics based on two central ideas - the idea of a *force* and the idea of *mass*. which are brought together in the central relation

$$F = ma$$

In principle, you can measure the mass of an object by applying a standard force to it (eg using a 'STANDARD' firework) and measuring its acceleration.

The greater the mass, the greater its *inertia* will be and hence the smaller the acceleration. Mass measured in this way should properly be called *inertial mass*.

But Newton did not stop at explaining all the laws of dynamics. Like Einstein after him, he also went on to explain the laws of gravity using the same fundamental concepts of *force* and *mass*. His second Big Idea was this: between two (small) masses *M* and *m* separated by a distance *r* there exists an attractive force *F* which is proportional to the product of the masses and inversely proportional to their separation. In symbols:

$$F = G\frac{Mm}{r^2}$$

where *G* is a constant equal to 6.67 x 10⁻¹¹ kg⁻¹m³s⁻².

This equation gives us an alternative and quite independent method of measuring mass. All we have to do is measure the force of gravity on the mass (eg by hanging it from a standard spring) in a standard gravitational field (eg the Earth's field.)

Mass measured in this way should properly be called *gravitational mass*.

Now it stands to reason (though it is neither obvious nor necessarily true) that if you strap two identical objects together, you will double both the inertial mass and the gravitational mass and it is this assumption that is at the heart of Galileo's argument. The heavy stone has twice the force of gravity on it - but then it *needs* twice the force in order to accelerate the same amount!

While it may stand to reason that two *identical* objects with the same inertial mass will have the same gravitational mass, there is absolutely nothing in Newton's theory which prevents two objects with the same inertial mass but made of different *materials*, from having *different* gravitational masses. It is conceivable, for example that *dense* materials like lead or gold would *weigh* more (or less) than their inertial mass would suggest.

'That's amazing! May be there are substances which we don't know about that have inertial mass but don't weigh anything at all!'

As far as Newton's theory is concerned, that is certainly possible but all the evidence we have to date strongly suggests that whatever substance an object is made of, inertial and gravitational mass are exactly the same. Even if there were tiny differences, we would notice it immediately. It would, for example, mean that the giant gaseous planets would orbit the Sun at a different rate from the smaller rocky ones (over and above the normal differences that is); it would mean that an aluminium satellite would have a different orbital period from the titanium shuttle which launches it; it would mean that the Earth would have a different orbit round the Sun than its own oceans which would, consequently, appear to fly off into space of their own accord! It is fairly evident that none of these things happen.

There is, therefore, plenty of evidence to show that the ratio of inertial mass to gravitational mass is the same for *all* materials, everywhere. But why? According to Newton's theory, there is no *reason* why, they just are. But according to Einstein, there is a very good reason: it is simply the Fundamental Principle of General Relativity - *The laws of Physics in a uniformly accelerated frame of reference are identical to the laws of physics in a uniform gravitational field.*

So just as the Fundamental Principle of Special Relativity explains the observed constancy of the speed of light, so the Fundamental Principle of General Relativity explains the observation that all objects fall with the same acceleration in a gravitational field. We can summarise this conclusion as follows:

Reassuringly Normal Consequence Number 15

Inertial mass and gravitational mass are one and the same thing.

But you haven't paid good money to the roller coaster owners just to discover Reassuringly Normal Consequences so, now that we have reached the summit and are ready to go, here is the next Bizarre one:

'That's ridiculous. Light can't bend. And I can tell you why too! When we say 'light travels in straight lines' we aren't stating an experimental fact; the way light travels defines what we mean by a straight line!'

Mmm... yes... you've got a good point there...

But before I have a chance to think up a suitable reply the catches on the roller coaster are suddenly released and we are plunging in free fall down into a gaping black hole.

Chapter 8 - The bending of light

This is easy to understand. Imagine that you are in a rocket playing with a laser pen. Suddenly, just at the moment when the Hitch-hikers Guide to the Galaxy shoots to the back of the rocket, you see the laser beam bend as well. Of course, the reason is obvious. The rocket is *accelerating* so that in the time it took for the light to cross the width of the rocket, the rocket had moved an extra distance causing the spot to hit the wall of the rocket further behind its original position.

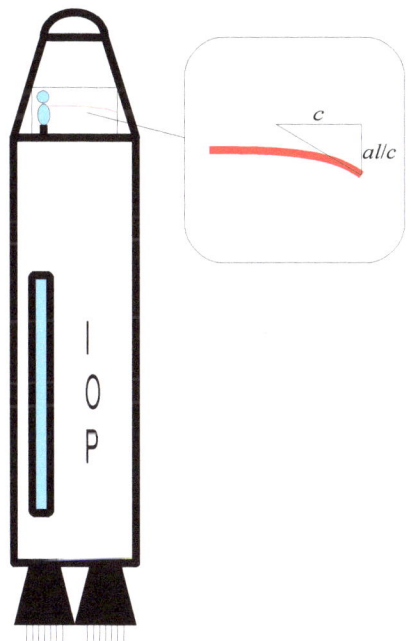

If the rocket has a width of l and is accelerating with an acceleration of a then in the time it takes for light to cross the width of the rocket t (where $t = l/c$) the rockets speed will have increased by $\delta v = at = al/c$. (It is here assumed that this increase is much smaller than the speed of light so we don't have to take any special relativity considerations into account)

The angle of deflection of the beam when it hits the wall (in radians) will therefore be approximately al/c^2.

Now from the Fundamental Principle, what is true in an accelerating system is true in a gravitational field so it follows that when a beam of light crosses a gravitational field of strength g ($= a$), it will bend through an angle α where:

$$\alpha = \frac{l\,g}{c^2}$$

It is also quite easy to show that the distance it will fall s is given by

$$s = \frac{1}{2}\frac{l^2\,g}{c^2}$$

Not surprisingly, this turns out to be rather small where the Earth's gravity is concerned. A laser beam shining horizontally over a distance of 1 km in the Earth's gravitational field falls by only 5×10^{-11}m. That is about a quarter of the diameter of an atom!

On the other hand, when light from a distant star passes close to the surface of the sun, it *is* significantly deflected and what is more, this deflection can fairly easily be measured. It causes the stars immediately behind the Sun to appear further away from the Sun than they really are.

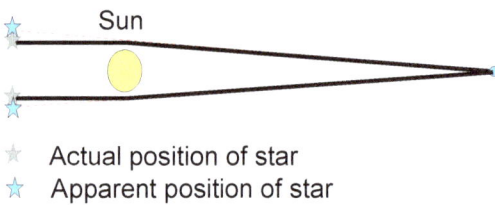

☆ Actual position of star
☆ Apparent position of star

The idea that gravity could bend light is not new, however. It would not have surprised Newton in the least who always thought light was a stream of particles which were influenced by gravity. On the other hand, the proponents of the wave theory of light, Huygens and, later, Maxwell would have hotly denied the possibility. After all, gravity acts on *mass*, how could it possibly affect an *electromagnetic wave*? And yet it does.

[I have to add a cautionary note here: in fact, gravity bends light by *more* than the amount calculated above. The flaw in our argument lies not in the approximation that the *speed* of the rocket is much less than the speed of light, but in the assumption that space and time are unaffected by the *acceleration* of the rocket (or, equivalently, the gravitational field). When the effect of the distortion of space and time is taken into account as well, the full General Theory predicts an angle of deflection of exactly twice the amount calculated above and the predictions of the full theory have been verified experimentally many times since Einstein first calculated the size of the effect.]

Prologue to chapter 9 - Down we go again

'Well except for that bit about the bending being twice what it should be which I didn't understand, that was pretty easy. Doesn't this roller coaster have anything a bit more exciting to offer?'

Well yes, it does - but first we have another rather familiar hill to descend. Remember - just hang on to that Principle!

As we cruise over the crest of the hill the roller coaster tilts over sideways and we find ourselves looking down a deep shaft. At the bottom of the shaft there are floodlights and workmen with clocks and measuring sticks - but they don't seem to be working very quickly - and their measuring sticks look very short - and, for some reason, they seem to be using red floodlights.

But all too soon the roller coaster rights itself and begins to hurtle down the next slope towards:

> ### Bizarre Consequence Number 17
> *Clocks go slow and rulers shrink at the bottom of a well.*

'Uh?'

Chapter 9 - Gravitational Time Dilation

Suppose you are playing with your red laser pen in the accelerated rocket but instead of shining it across the rocket, you shine it straight towards the front of the rocket. What, if anything, will happen to the light then?

'I know the answer to that one! It will just go straight forwards at the speed of light!'

Well done! You have obviously learned your lessons well. But something happens, all the same.

'What?'

Its *colour* changes. You see, by the time the light has travelled a distance h up the rocket (relative to you in the rocket, that is) the rocket has increased its speed by $\delta v = ah/c$. What this means is that, in effect, the receiver at the front of the rocket is continually moving *away* from the source with speed ah/c. Now if we restrict ourselves to small heights and small accelerations (ie if $ah \ll c$) this produces a Doppler shift equal to:

$$\lambda = \lambda_0(1 + ah/c^2)$$

Now we know that the effects of gravity are exactly the same as the effects of acceleration so we must conclude that when light climbs up through a gravitational field, its wavelength increases according to the formula:

$$\lambda = \lambda_0(1 + gh/c^2)$$

where g is the gravitational field strength (assumed uniform over the distance h)

I am sure you know that the gravitational potential energy of a mass m lifted a height h in a gravitational field of strength g is mgh. You may also know that *gravitational potential* (ϕ) is defined as the gravitational potential energy per unit mass. It follows that when you move up a distance h through a gravitational field g, the change in gravitational potential $\delta\phi$ is equal to

$$\delta\phi = \frac{mgh}{m} = gh$$

This means that we can write our equation for the Doppler shift more generally like this:

$$\lambda = \lambda_0(1 + \delta\phi/c^2)$$

The greater the gravitational potential difference which the light has to climb, the greater the Doppler shift which is produced. On the other hand, because of the c^2 term, the shift is very small indeed. The gravitational potential at the surface of a star of radius R and mass M is equal to:

$$\phi = -\frac{GM}{R}$$

(The minus sign is necessary because the gravitational potential is defined to be zero at an infinite distance from the star. At the surface it is *less* than zero. For the proof of this formula see the appendix)

In rising from the star, a beam of light passes up through a change in gravitational potential equal to:

$$\delta\phi = \frac{GM}{R}$$

so the formula for the gravitational red shift experienced by a beam of light emitted from the surface of a star of mass M and radius R is:

$$\lambda = \lambda_0\left(1 + \frac{GM}{Rc^2}\right)$$

Now for the Sun, $M = 2 \times 10^{30}$ kg, $R = 7 \times 10^8$ m so the expression GM/Rc^2 computes to 2×10^{-6}. A Doppler shift of less than one part in a million is very small and corresponds to a recession speed of only 600 ms^{-1} - much less than the Doppler shift in wavelength induced by the thermal motion of atoms on the surface of the Sun, which makes the effect difficult to detect. Nevertheless, the effect is real and in the case of more massive stars, it should not be ignored. What is more important at the moment is the implication this analysis has on the rate of passage of time at the two ends of the rocket.

Suppose that instead of shining a laser pen towards the front of the rocket, you use it to send timing signals to a fellow astronaut situated at the front. If. by your watch, you send timing signals every T_0 seconds, these will be Doppler shifted just like the light beam and your friend will receive timing signals every T seconds where:

$$T = T_0(1 + \delta\,\phi/c^2)$$

What this means is that the clock at the back of the (accelerating) rocket goes slower than the clock at the front. It also means that a clock at the bottom of a well goes slower than a clock at the top. What is more, this effect has actually been demonstrated! Unbelievable though it may seem, the effect was observed at Havard university in 1960 over a distance of 22.5 m. (if you are wondering why such a strange distance it is because that is the height of the stair well at the Jefferson Physical Laboratory!) A quick calculation shows that the Doppler shift over this distance is equal to 2.5×10^{-15}. What the Havard scientists, Pound and Rebka, did was to build a gamma ray source and a gamma ray detector so finely tuned that if the frequency of the source changed by only this amount, the detector would reject the signal. The source and the detector were placed together at the bottom of the shaft and matched exactly. Then the detector was taken upstairs and, as expected, the signal was rejected. To confirm the effect absolutely, the source was raised at a steady speed towards the detector. The blue doppler shift caused by the speed was just sufficient to counteract the red shift caused by the gravity field and the detector sprang into life again. What was this speed? 2×10^{-15} times the speed of light, of course. You want me to calculate it for you? All right. It works out to be 2.7 millimetres *per hour*!

The effect is also measurable using atomic clocks in a high flying aircraft. You may recall that the velocity time dilation effect of a jet plane flying at 200 ms^{-1} for 10 hours was 8 ns *slow*. Now suppose that the plane was flying at a height of 10,000 m during this time. The change in gravitational potential between this height and the ground is approximately $gh = 10 \times 10,000 = 100,000$ J kg^{-1} so the clock in the plane will run *faster* than the clock on the ground by a factor of gh/c^2. For a period of 10 hours, this works out to be 40 ns, so the effect of gravitational time dilation in this case is actually larger, and in the opposite sense, to the effect of velocity time dilation.

The derivation of the gravitational time dilation formula presented so far has two disadvantages. Firstly, I have used an approximate formula for the Doppler shift only applicable for small values of h and a. More seriously though, it ignores the second order effects of special relativity due to the fact that the increase in speed of the rocket during the time t (whose time t anyway?) is not exactly equal to at. In order to generate a formula which is applicable to situations in which ah is large, we need to look at another way of producing artificial gravity.

Science fiction writers are fond of describing large futuristic space stations as a large spoked wheel, spinning slowly on its axis.

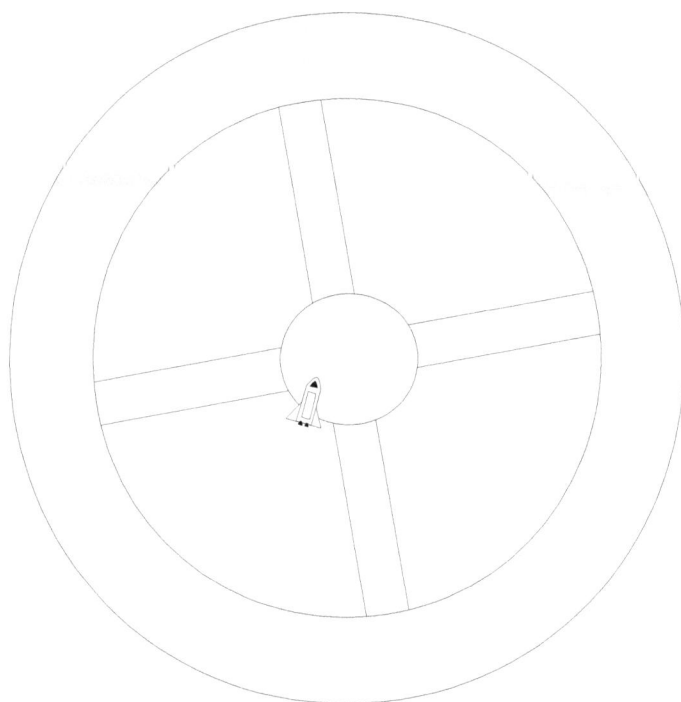

As the station spins, astronauts at the periphery subjectively feel an outward force, popularly (and correctly) known as the centrifugal force which feels to them just like gravity. From the point of view of an observer in a shuttle craft waiting to dock in the centre of the station, we can see that what is in fact happening is that the metal structure of the station is exerting an *inward* force on the astronauts which is giving them an *inwards centripetal* acceleration causing them to move in a circle.

Correction. I am not allowed to say that that is what is *in fact* happening. General Relativity tells us that the astronauts point of view is just as valid as ours. So what does the universe really look like from the point of view of the astronauts on board the space station? To them, the station is, of course, stationary. But the station is generating a rather strange form of gravity which is zero at the centre of the station and which *increases* steadily as you move towards the rim. Moreover, the force of gravity is *outwards*, not inwards.

Now you know that the centripetal acceleration of a mass m rotating round in a circle of radius r with angular speed ω is equal to $r\omega^2$ so the strength of gravity at the rim will be:

$$g = R\omega^2$$

(where R is the radius of the station) and the gravitational potential difference between the centre and the rim will be given by:

$$\Delta\phi = \int_0^R r\omega^2 \, dr = \tfrac{1}{2}R^2\omega^2$$

What about clocks placed at the centre and at the rim? From our point of view outside the station, we can see that the clock at the rim is going more *slowly* than the clock at the centre because the clock at the rim is moving while the clock at the centre is not. In fact we can write down the relation between the time as measured by the rim clock T_{rim} in terms of the time measured by the clock at the centre T_{centre} as follows using the special relativity time dilation formula:

$$T_{rim} = \frac{1}{\sqrt{1-v^2/c^2}} T_{centre} = \frac{1}{\sqrt{1-R^2\omega^2/c^2}} T_{centre}$$

More generally, by putting $R^2\omega^2 = 2\Delta\phi$ we can write down the general formula for the relativistic time dilation between two points where a gravitational potential difference exists as follows:

$$T = \frac{1}{\sqrt{1-2\Delta\phi/c^2}} T_0$$

(If you compare this formula with the similar formula on page 58 you will see that the later is simply the binomial expansion of this one approximated to the second term.)

Now the astronauts on board the space station will observe exactly the same time dilation effect as we do - it is just that they will explain it differently.

(It is very important to note that, like velocity time dilation, gravitational time dilation is a *relative* phenomena which depends not on the local strength of gravity, but on the *difference in gravitational potential* between two points. At the very centre of the Earth, for example, there is no gravity - but clocks will still run more slowly there.)

So much for clocks - what about rulers? This is where things start to get difficult. Suppose that from our stationary birds eye view in the shuttle, we watch one of the astronauts measuring out the radius of the space station with a metre ruler. Since the ruler is always at right angles to its direction of motion, we can see that it remains the same length and his measurement of the radius agrees with ours namely R.

But what happens when he measures the circumference? We see that his ruler is contracted because of the speed and so, instead of getting the expected value of $2pR$, he ends up with an answer that is *bigger* than this! What is he to make of this? Does gravity alter the fundamental mathematical constants like π? No, I think that is going too far. What we can say is that gravity distorts the structure of space in such a way that, in a non-uniform gravitational field, the circumference of a circle is no longer equal to 2π times the radius. In fact, this is such an important conclusion, it deserves a box of its own.

Bizarre Consequence Number 18

In a non-uniform gravitational field, space is non-Euclidean; the circumference of a unit circle is not equal to 2π and the angles of a triangle no longer add up to 180°.

'So what happens to rulers in a gravitational field? Do rulers shrink or what?'

Well I find it difficult to give a straight answer to that question. It is true that if I measure the radius and the circumference of a large star with a metre ruler (!) I will find that the circumference is *less* than 2π times the radius. (The reason why it is *less* not *more* is that real gravity is an attractive force, unlike the centrifugal effect inside the space station which is an outward force.) This would appear to suggest that as you take the ruler down to deeper and deeper depths inside the star - ie to lower and lower gravitational potentials - the ruler shrinks. And because gravitational potential is a *scalar* quantity, the ruler has to shrink in *all directions*, not just parallel or perpendicular to the local gravitational field. However, if a ruler shrinks in *all* directions, what sense does it make to say that it shrinks at all? The truth is that the effect of gravity is far more subtle than just a question of shrinking rulers. No, gravity actually distorts the space in which the rulers exist. But to go any further along that road requires the use of mathematical concepts far beyond my competence.

Prologue to chapter 10 - Oblivion

'Wow! That was a bit weird' you shout as the roller coaster shoots up to the top of the next hill. 'Where do we go now?' you ask.

The roller coaster track ahead of us goes horizontally for a few meters and then - just stops!

As the train reaches the end of the track it tilts over 90^0 and, whether time actually stops still or whether it is just our perception of it that stops, I do not know, but for an instant we are suspended over the most terrifying sight. Below us is a deep, interminable shaft. Along the sides of the shaft are electric lights at regular intervals but as we look down the shaft, the lights in the distance get redder and redder until they fade from view completely. At the bottom of the shaft - no, there is no bottom - there is just an aching blackness.

Suddenly we are in free fall; the lights are flashing past us faster and faster; as we look up, the shining blue sky which we left behind is turning first violet, then ultra violet, and now begins to bathe us in X-rays and even γ-rays; and then the aching blackness which we saw from above appears to grow in size until it is all around us, even above us. The bright patch of gamma-rays which was our home on Earth is vanishing fast and suddenly, it is gone. We are completely alone in the darkness with nothing to tell us where we are or how fast we are travelling except that we begin to experience a very unpleasant sensation of being stretched and squashed at the same time. Rapidly the stretching and squashing increases to the point where our bodies are dismembered and pulled into long thin strands, but we are past pain - we have passed into oblivion . . .

Chapter 10 - Black Holes

In 1783, the Rector of Thornhill parish church in Yorkshire, a man called John Michell wrote to the Royal Society with an extraordinary idea. Newton had showed that for every planet, there was a critical speed called the escape velocity which it was necessary to achieve if you were to throw a rock off the surface of the planet into space and it is easy to show that, for a planet (or star) of mass M and radius R, the escape velocity v is given by the formula:

$$v_{escape} = \sqrt{\frac{2GM}{R}}$$

Michell's idea was this. What if a star was either so massive, or alternatively, so small that the escape velocity was equal or greater than that of light? According to the corpuscular theory of light which was still popular at that time, at least in England, light would never be able to escape from such a star and even though it was burning brightly, it would look to us completely black. The radius of such a 'black star' would be equal to:

$$R = \frac{2GM}{c^2}$$

The idea did not catch on. Within a couple of decades, Thomas Young had shown pretty conclusively that light was in fact a wave and therefore probably immune to gravity. Nevertheless, we have now shown that light is indeed susceptible to gravity and that, as it climbs away from the surface of a star, it is not slowed down - it is red-shifted instead.

The modern equivalent of Michell's idea is therefore this. Is it possible for a star to be so massive that light is red-shifted all the way to infinite wavelength (or zero frequency)? In which case it would become completely invisible just like Michell's corpuscles. To answer this question we must use the accurate formula for gravitational time dilation:

$$T = \frac{1}{\sqrt{1-2\Delta\phi/c^2}} T_0$$

and ask ourselves under what circumstances will T (the observed rate of flow of time on the surface of the star) to become infinite. The answer is of course when:

$$1-2\Delta\phi/c^2 = 0$$

$$2\Delta\phi/c^2 = 1$$

Now the gravitational potential at the surface of a star is (see Appendix H)

$$\phi = -\frac{GM}{R}$$

so putting $\Delta\phi = -\phi$, we get

$$\frac{2GM}{Rc^2} = 1$$

and hence

$$R_s = \frac{2GM}{c^2}$$

which happens to be exactly the same formula that Michell proposed in the first place!

This is the radius of a Black Hole - the so called Schwarzschild radius. (As we have seen, the actual radius, eg the distance from the centre to the edge, is in general larger than this and for a black hole it is probably infinite. The radius used here is defined as being the circumference divided by 2π.)

54

For a black hole with the same mass as the Sun, R_s turns out to be almost exactly 3 km. It is clear that our own Sun is a long way from being a black hole but theoretical physicists and astronomers are now fairly convinced that most stars with a mass greater than about 5 solar masses will end their lives by turning into a black hole. Moreover the existence of supermassive black holes at the centres of galaxies is widely accepted and the existence of tiny mini-black holes is a possibility.

What would the mass of a black hole the size of an atom be? The answer is an incredible 10^{17} kg which is about the mass of an iron meteorite 32 km across. If the Earth were to encounter one of these it would either cause a massive explosion big enough to obliterate all life on Earth, or. more likely, it would drill a neat hole through the Earth and come out the other side!

The gravitational field strength at the surface of a black hole (or the Event Horizon as it is more properly called) is equal to

$$g_s \;=\; \frac{GM}{R_s^2} \;=\; \frac{c^2}{2R_s}$$

What this means is that really big black holes have very modest gravitational field strengths at their event horizons. For example, a black hole with a mass of 1.5 trillion solar masses (that is about a thousand galaxies) would have a Schwartzchild radius of about half a light year and a gravitational field strength at the surface of 10 ms^{-2} – the same as Earth. If you wandered close to or even inside this black hole, you would not feel anything that you do not feel every day here on Earth – but, of course, you wouldn't be able to get out, however hard you tried!

It is instructive, when considering the possibility of black holes of varying sizes to consider the density of the material inside the hole. (For our purposes here we shall assume that the hole is 'filled' with a material on uniform density up to its event horizon.) Putting

$$M \;=\; \frac{4}{3}\pi R_s^3 \rho$$

we obtain

$$R_s \;=\; \frac{8\pi R^3 \rho G}{3c^2}$$

so

$$R_s \;=\; \sqrt{\frac{3c^2}{8\pi \rho G}} \;=\; \frac{1.26\times 10^{13}}{\sqrt{\rho}}$$

What this means is that the smallest black hole which can be made with ordinary matter (eg iron of density 7000 kg m^{-3}) will have a radius of 1.5 x 10^{11} m. By coincidence, this happens to be exactly the same as the orbital radius of the Earth. So imagine the whole of the solar system out to the Earth filled with iron and you have a simple black hole. It would have a mass of 51 million Suns and a gravitational field at the surface of 30,000 g!

Actually this scenario is quite academic because, as I have said earlier, it only needs about 5 solar masses of ordinary matter to make a black hole owing to the inability of ordinary matter to withstand the crushing effect of gravity. On the other hand, if we choose to make a black hole out of very rarefied matter, then the 'pressure' at the centre does not have to be very great at all. In fact, we could be living inside a black hole right at this very moment! If so, the last formula quoted above could be telling us something very important about the relation between the amount of matter in the universe and the density of material that it contains.

Current theories about the size of the universe suggest that it might have a 'radius' of about 15 billion light years or 1.4 x 10^{26} m. Plugging this figure into the formula gives an average density for the universe of about 8 x 10^{-27} kg m^{-3}. Now the mass of a single hydrogen atom is 1.7 x 10^{-27} kg, so this density corresponds to about 5 hydrogen atoms per cubic metre.

It is exceedingly difficult to estimate the average density of the universe but the best present day estimates of the average density of all the observable matter in the stars and galaxies is about 0.3 x 10^{-27} kg m^{-3}, that is one thirtieth of the required value. When you consider by what disparate routes the two figures were arrived at in the first place it is surprising that it comes anywhere close at all. What it means is this. If the average density of the universe is *less* than the critical density, we are living in what is known as an *open* universe. The universe is expanding now and will continue to expand for ever. If on the other hand, our estimates are wrong and/or we discover other sorts of matter which we have not included so far and the density of the universe is greater than the critical density, then we are effectively living inside a black hole; the universe will one day stop expanding and will fall back in on itself.

The most satisfying result would be that the universe has exactly the right density to make it into a black hole - not an atom more, not an atom less. If so, there will have to be a reason for it just as there was a reason why the speed of light is invariably constant and why inertial mass is exactly the same as gravitational mass.

Or perhaps there isn't a *reason* why at all. Perhaps we should follow Einstein's example and simply put it in a box and call it give it an impressive name! How about this?

The Fundamental Principle of Cosmology

Every universe is inside a black hole and inside every black hole is a universe.

Now there's a thought!

Epilogue

. . . but just when we can stand it no more, the stretching and squashing begin to ease and a violet light appears ahead of us which appears to spread wider and wider turning bluer and bluer. The electric lights appear again but they are not like our lights. The aching blackness recedes and suddenly we tumble out into a graceful landscape with a blue sky, green grass and distant purple-headed hills.

*'What the **** happened then?'* you say. (You always were prone to colourful language.)

'I think we have fallen through a wormhole into another universe.' I reply.

It is true that there is something strange and unfamiliar about everything. The grass has a strange feel and the sky is not quite the right colour. I pick up a stone and let it drop *'Well the laws of Physics seem to be much the same here.'* As we look round we find that we are standing on the edge of a vertical shaft. Peering down we see the electric lights reddening in the distance and that aching void again.

Turning suddenly, we see a strange figure standing nearby with an outstretched hand.

"Welcome" he says, "we have been expecting you to drop in for some time now . . ."

Appendix A - The 1g rocket problem

Consider a rocket accelerating from rest at a constant acceleration of 1g. What do we actually mean by this, taking relativistic effects into account? Obviously the rocket cannot go on increasing its speed at a constant rate (from the point of view of someone who remains at rest) because it would eventually travel faster than light. What happens is that the rocket approaches the speed of light exponentially like this.

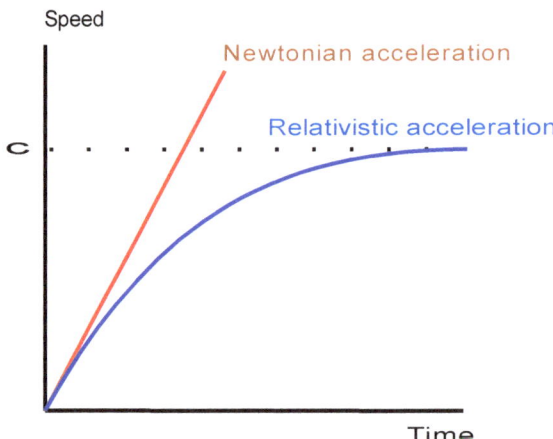

So what do we mean by *constant acceleration*? Well, from the point of view of the astronauts on board, what we mean is that they experience a constant artificial gravity field which is produced by the steady thrust of the continuously firing rocket motors which feels exactly like 1g back on Earth. To them, the spacecraft is permanently stationary but the universe outside the ship appears to be passing by faster and faster in the same way as in the illustration above.

Now it might appear that we need to use the ideas of General Relativity here, but that is not necessary. We just need to define carefully what we mean by this sort of acceleration. In a Newtonian world, if an object travelling at a speed v accelerates with an acceleration a for a small time δt, the final speed is given by

$$v + \delta v = v + a\delta t$$

In a relativistic world, however , we must use the relativistic formula for the addition of velocities (see appendix *), ie:

$$v + \delta v = \frac{v + a\delta t}{1 + v a\delta t / c^2}$$

This leads to

$$(v + \delta v)(1 + v a\delta t / c^2) = v + a\delta t$$

$$v + v^2 a\delta t / c^2 + \delta v + v a\delta v . \delta t / c^2 = v + a\delta t$$

The two v's cancel - and we can also cross out the term which contains two d terms because in the limit, this term will be much smaller than the single d terms so:

$$\delta v = a\delta t - v^2 a\delta t/c^2$$

$$\delta v = a\delta t(1-v^2/c^2)$$

Now all we have to do is to integrate this expression to find out how v varies with t.

First separate the variables:

$$\frac{\delta v}{(1-v^2/c^2)} = a\delta t$$

$$\frac{\delta v}{(c^2-v^2)} = \frac{a}{c^2}\delta t$$

Now integrate:

$$\int \frac{1}{(c^2-v^2)}dv = \frac{a}{c^2}t$$

Fortunately this is a standard integral which my maths book tells me is:

$$\frac{1}{c}\tanh^{-1}\frac{v}{c} = \frac{a}{c^2}t$$

(Since $v = 0$ at $t = 0$, there is no constant of integration). A bit of simple manipulation leads us to the first result:

$$v = c\tanh\frac{a}{c}t$$

NB the time t referred to here is the *integral* of the proper time, which is the *journey* time - ie the time as experienced on board ship. So this formula tells you how fast you will be going after you have been travelling for a time t.

Next it would be nice to know how far you get in this time. To do this we must take into account the fact that the faster you go, the more contracted the miles that pass by are! In a Newtonian world the distance δs travelled in a short time δt at a speed v is:

$$\delta s = v\delta t$$

but in a relativistic world, you actually get a lot further because all the distances outside your ship are length contracted by a factor of γ so:

59

$$\delta s = \gamma v \delta t$$

$$\delta s = \frac{v}{\sqrt{1 - v^2/c^2}} \delta t$$

Substituting our formula for the speed of our ship after a time t we get:

$$\delta s = \frac{c \tanh at/c}{\sqrt{1 - (\tanh^2 at/c)}} \delta t$$

Now fortunately, $1 - \tanh^2 x = \operatorname{sech}^2 x$, and of course $\tanh x = \sinh x / \cosh x$ so this horrendous expression simplifies down to just:

$$\delta s = c \sinh \frac{at}{c} \delta t$$

which even I can integrate!

$$s = \frac{c^2}{a} \cosh \frac{at}{c} + K$$

We can't ignore the constant of integration this time because at $t = 0$, $s = 0$ but $\cosh(0)$ is 1 not zero. The final result is therefore:

$$s = \frac{c^2}{a} \left(\cosh \frac{at}{c} - 1 \right)$$

The third thing we should like to know is how old will our friends be when we get back! Well, for every second we travel at a speed v our stay-at-home friends will age g seconds ie:

$$\delta t_{home} = \frac{\delta t}{\sqrt{1 - v^2/c^2}}$$

$$t_{home} = \int \frac{1}{\sqrt{1 - \tanh^2 at/c}} dt$$

$$t_{home} = \int \cosh \frac{at}{c} dt$$

from which we obtain:

$$t_{home} = \frac{c}{a} \sinh \frac{at}{c}$$

(the constant of integration is zero because sinh(0) = 0 as required)

I find it particularly pleasing that the solution to the problem of the 1g rocket has such simple and elegant answers, particularly when we work in years and light years. In these units, both a and c are equal to 1. If we think of a round trip which is to take T years (as measured on board), the outward journey will consist of two phases, an acceleration for $T/4$ years, and a deceleration phase of the same length. The return journey will be just the same.

The maximum speed reached will be:

$$V = \tanh \frac{T}{4}$$

The distance you get to will be:

$$S = 2\left(\cosh \frac{T}{4} - 1 \right)$$

and the age of your newborn baby son when you get back will be:

$$T_{home} = 4 \sinh \frac{T}{4}$$

To give you an idea of what these expressions look like in practice, here is a table of results:

Journey time (years)	Maximum speed reached (% of light)	Distance reached (light years)	Age of your baby son when you get back (years)
0	0	0	0
1	24	0.1	1
2	46	0.3	2
3	64	1	3
4	76	1	5
5	85	2	6
6	91	3	9
7	94	4	11
8	96	6	15
9	98	8	19
10	99	10	24

It is worth noting that a return journey to a star 10 light years away would only take 10 years of astronaut time (24 years back at home). If you are prepared to journey for 20, 30 or 40 years, you could get to stars 150, 1800 and 22,000 ly away respectively (that is a quarter of the way across the galaxy!) and if you are not bothered about coming home you could travel to the edge of the known universe in a mere 47 years (although whether there would be any universe left by the time you got there is another matter!).

Appendix B - Length Contraction.

Using the context of the river race. if we see Albert and Beatrice arrive back at exactly the same instant and *knowing that Beatrice cannot row faster than Albert*, we must conclude that Beatrice has a shorter distance to row.

How much shorter? Suppose that Albert rows a distance l_A and Beatrice rows l_B.

$$\text{Time for Albert to finish} \ = \ \frac{2l_A}{\sqrt{c^2-v^2}}$$

$$\text{Time for Beatrice to finish} \ = \ \frac{l_B}{c-v} \ + \ \frac{l_B}{c+v} \ = \ \frac{2l_B c}{c^2-v^2}$$

Now these must be equal so:

$$\frac{2l_a}{\sqrt{c^2-v^2}} \ = \ \frac{2l_b c}{c^2-v^2}$$

from which we get:

$$l_b \ = \ \sqrt{1-v^2/c^2} \ l_A$$

What this means is that from the point of view of a stationary observer, a rod of *proper length l_0* (ie whose length is l_0 when at rest) will have a length l given by

$$l \ = \ \sqrt{1-v^2/c^2} \ l_0 \ = \ l_0/\gamma$$

when it moves with a velocity v in a direction parallel to its length. (It will always appear to have the same length l_0 perpendicular to the direction of motion.)

Appendix C - Time on a distant star

How would you synchronise two clocks which are at rest but in different places?

One way to do it would be this:

- measure the distance *l* between the clocks

- set clock A to zero (but do not start it)

- set clock B to read time *l/c* (but again do not start it)

- start clock A and simultaneously send a pulse of light towards clock B

- when the pulse of light reaches B, start clock B

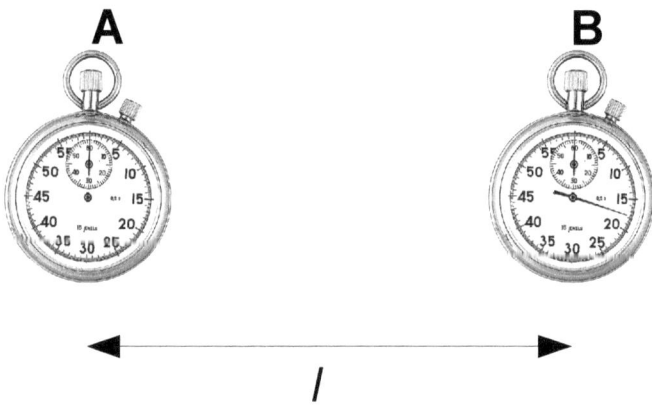

The two clocks will now be synchronised, at least to an observer at rest with respect to the two clocks.

But what if I were to watch you synchronising a pair of clocks like this while you sailed by me (in the direction B to A) at a speed *v*? I would disagree about your calculation in two ways. Firstly the distance between the clocks would be contracted by a factor of γ. Secondly, while the light beam was travelling from A to B at what is to me a constant speed *c*, B would be racing towards the light beam at a speed *v*. To me, the light beam would be gaining on the clock at a relative speed of *c + v* so the time taken for the beam to get from A to B would be:

$$T = \frac{\sqrt{1-v^2/c^2} \cdot l}{(c+v)}$$

This can be rearranged and simplified as follows:

$$T = \frac{l}{c} \frac{\sqrt{c^2-v^2}}{(c+v)}$$

$$T = \frac{l}{c}\sqrt{\frac{(c+v)(c-v)}{(c+v)(c+v)}}$$

$$T = \frac{l}{c}\sqrt{\frac{c-v}{c+v}}$$

What this means is that while you set clock B to read l/c, I think you should have set it to the *smaller* time above. To put this another way, it seems to me that all your supposedly synchronised clocks ahead of you are running behind the true time and all those behind you are running ahead of the true time.

Now the *difference* in the two times is equal to:

$$\Delta T = \frac{l}{c} - \frac{l}{c}\sqrt{\frac{c-v}{c+v}}$$

We can simplify this formula considerably if we assume that v is a lot smaller than c. First we rewrite the formula as follows:

$$\Delta T = \frac{l}{c}\left(1 - (1-v/c)^{1/2}(1+v/c)^{-1/2}\right)$$

hence using the Binomial Theorem we get:

$$\Delta T = \frac{l}{c}\left(1 - (1-\tfrac{1}{2}v/c+...)(1-\tfrac{1}{2}v/c+...)\right)$$

from which, by rejecting all second order terms we arrive at:

$$\Delta T = \frac{lv}{c^2}$$

Appendix D - The addition of speeds

Consider a spaceship A of length l_A travelling past you at a speed v_A passing a second spaceship B of length l_B travelling in the opposite direction at a speed v_B (both speeds measured with reference to you). The question we need to answer is what speed v_X does B appear to be going past the astronauts in A?

We need to consider two vital events; first contact, when the nose cones of the two spaceships meet and second contact when the two tails part.

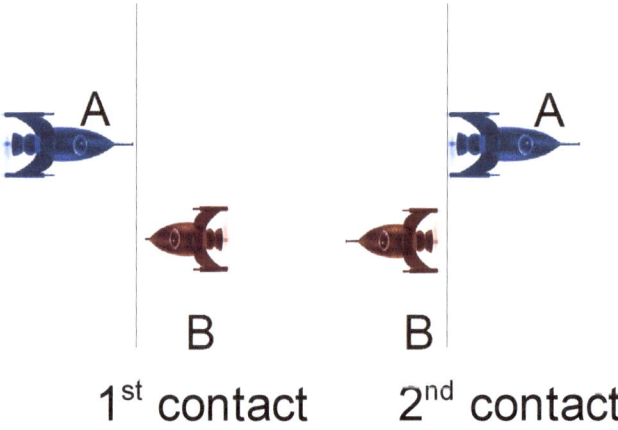

What is more, let us suppose that these two events occur *at the same place* from your point of view. This means that the spaceships have to be just the right length so that they both pass you in the same time. Note that from your point of view, both ships are contracted by the factors γ_A and γ_B respectively.

Now what does the commander of ship A see? He sees *you* travelling past at a speed v_A and also ship B travelling past at some greater speed v_X.

Note that to the occupants of ship A, B is length contracted by a factor γ_X.

Now commander A can calculate the time between first and second contact in two ways. First he sees you travelling a distance l_A at a speed v_A. Secondly, he sees ship B (whose length is contracted to l_B/γ_X) travel a distance equal to $l_A + l_B/\gamma_X$ at a speed v_X. It follows that:

$$\frac{l_A}{v_A} = \frac{(l_A + l_B/\gamma_X)}{v_X}$$

$$l_A v_X = l_A v_A + l_B v_A/\gamma_X$$

$$l_A(v_X - v_A) = l_A v_B/\gamma_X$$

An identical argument from commander B's point of view leads to:

$$l_B(v_X - v_B) = l_B v_A / \gamma_X$$

(It should be pointed out that because of the invariance of velocity, if commander A sees ship B moving at a speed v_X then commander B will see ship A moving at exactly the same speed and contracted by the same γ factor.)

Now multiplying the two equations together eliminates the lengths of the two ships and leaves us with a relation between the three velocities. The rest is just algebra.

$$(v_X - v_A) \cdot (v_X - v_B) = v_A v_B / \gamma_X^2$$

Now

$$\gamma_X = \frac{i}{\sqrt{1 - v_x^2 / c^2}}$$

so:

$$1/\gamma^2 = 1 - v_X^2 / c^2 = (c^2 - v_X^2)/c^2$$

hence:

$$(v_X - v_A) \cdot (v_X - v_B) = v_A v_B (c^2 - v_X^2)/c^2$$

from which we obtain by straightforward algebra the result we desire:

$$v_X = \frac{v_A + v_B}{1 + v_A v_B / c^2}$$

Appendix E - Travelling 'faster than light'

If you travel for a distance x at a speed v, owing to length contraction, the proper time interval (ie the number of years you age during the journey) between setting out and arriving will be:

$$T = \frac{x}{\gamma v} = \frac{\sqrt{1-v^2/c^2}}{v} \cdot x = \frac{\sqrt{c^2-v^2}}{cv} \cdot x$$

$$T = \sqrt{c^2/v^2-1} \cdot \frac{x}{c}$$

A light beam, on the other hand will actually take:

$$T = \frac{x}{c}$$

If we put these two expression equal, we can find out at what speed it is necessary to travel in order to get the effect of travelling as fast as light.

$$\sqrt{c^2/v^2-1} \cdot \frac{x}{c} = \frac{x}{c}$$

$$\sqrt{c^2/v^2-1} = 1$$

$$c^2/v^2-1 = 1$$

$$c^2 = 2v^2$$

$$v = \frac{c}{\sqrt{2}}$$

ie 71% of the speed of light.

Let me just say again what this means. You are not actually travelling faster than light – but you will reach Alpha Centauri, 4 light years away, in only 4 of *your* years.

Appendix F - Relativistic Kinetic Energy

Suppose a mass of rest-mass M_0 is accelerated from rest by a constant force F for a time t.

Under Newtonian mechanics, the acceleration of the mass a will be

$$a \ = \ \frac{f}{M_0}$$

the distance travelled s will be

$$s \ = \ \tfrac{1}{2}at^2 \ = \ \frac{Ft^2}{2M_0}$$

and the final speed reached will be

$$v \ = \ at \ = \ \frac{Ft}{M_0}$$

from which we deduce that

$$Ft \ = \ M_0 v$$

Now the Kinetic Energy KE acquired by the mass will be equal to the work done by the force which is, of course, force × distance or Fs. Hence:

$$KE \ = \ Fs \ = \ \frac{F^2 t^2}{2M_0} \ = \ \frac{M_0^2 v^2}{2M_0}$$

$$KE \ - \ \tfrac{1}{2}M_0 v^2$$

Now in order to carry out the same kind of analysis using Special Relativity, we need to use the results obtained during the analysis of the constantly accelerated rocket, namely:

$$s = \frac{c^2}{a}\left(\cosh\frac{at}{c} - 1\right)$$

and

$$v = c \tanh \frac{a}{c} t$$

What we need to do is eliminate t from these two equations which can be done using the following standard relation:

$$\frac{i}{\cosh^2 \theta} + \tanh^2 \theta = 1$$

The algebra is a bit messy but it is not difficult and reduces to

$$as = \frac{c^2}{\sqrt{1-v^2/c^2}} - c^2 = c^2(\gamma-1)$$

Now we need to think carefully what we have found out. This equation gives us a relation between the speed reached and the distance travelled for a rocket undergoing a constant acceleration of a. But from whose point of view? The person on the rocket or the person who remains at rest?

Well there is no problem about the speed v. As we have seen, both observers agree about the relative speed of any object. But in any case, it is the speed as seen by the *stationary* observer which we are talking about

The distance s is the distance travelled as seen by the *stationary* observer as well (remember it is the actual distance travelled to the star, not the length contracted distance).

What about a the acceleration? This is the acceleration *as experienced by the occupants of the rocket*. You will remember that the rocket is supposed to accelerate in such a way that the occupants of the rocket experience a constant 'artificial gravity'. As far as the astronauts are concerned, the rocket motors produce a constant thrust of F and the rocket has a constant mass of M_0 so we can still assume that

$$a = \frac{F}{M_0}$$

hence

$$Fs = c^2(\gamma-1)$$

Now normally we would just write

$$KE_r = Fs$$

but this requires a bit of justification. You see, s is the distance travelled as seen by the stationary observer while F is the thrust as measured by the astronauts and it is not immediately obvious that you can multiply these two quantities together.

Suppose that instead of accelerating a rocket with on board rocket motors, we consider accelerating an electron using a (stationary) electric field E.

The question is; is the force *as experienced by the electron* the same as the force *exerted on the electron by the accelerator*? In a sense, there is no answer to this question because it all depends on what you mean by force. What we are effectively doing is *constructing* a relativistic definition of force which is as close to the Newtonian definition as possible. One of the things which we would like our relativistic force to do is to obey Newton's third law which can be stated as action and reaction are equal and opposite; so if we assume that this is true we can go ahead and complete the proof. ie:

$$KE_r \; = \; M_0 c^2 (\gamma - 1)$$

Appendix G - The relation between energy and momentum

The total relativistic energy E and the relativistic momentum p of a body are given by the following expressions:

$$E = \gamma M_0 c^2 = \frac{M_0 c^2}{\sqrt{1-v^2/c^2}}$$

$$p = \gamma M_0 v = \frac{M_0 v}{\sqrt{1-v^2/c^2}}$$

We wish to eliminate v from these equations.

First square and multiply across:

$$E^2(1-v^2/c^2) = M_0^2 c^4$$

$$p^2(1-v^2/c^2) = M_0^2 v^2$$

Now for a diabolically cunning move, multiply the second equation by c^2 and subtract!

$$(E^2 - p^2 c^2)(1-v^2/c^2) = M_0^2 c^4 (1-v^2/c^2)$$

from which we obtain:

$$E^2 - p^2 c^2 = M_0^2 c^4$$

An alternative (and in my opinion better) way of writing this equation is:

$$E^2 - E_0^2 = p^2 c^2$$

where E_0 is the rest-mass energy of the body.

It is instructive to compare this expression with the non-relativistic relation between energy and momentum which is calculated as follows

$$KE = \tfrac{1}{2}Mv^2 \quad \text{and} \quad p = Mv$$

$$\text{so} \quad KE = \frac{p^2}{2M}$$

It is not easy to see, at first, how the relativistic expression will reduce (as it must) to the non-relativistic one when v is small, but it does. Watch!

Since

$$E^2 - E_0^2 = p^2 c^2$$

we can write

$$(E - E_0)(E + E_0) = p^2 c^2$$

Now $(E - E_0)$ is just the relativistic kinetic energy KE, which, at low speeds approximates to the ordinary kinetic energy KE.

At low speeds, the total relativistic energy E and the rest-mass energy E_0 are virtually equal and equal to Mc^2 so:

$$KE \,.\, 2M c^2 = p^2 c^2$$

from which it is easy to see that

$$KE = \frac{p^2}{2M}$$

as expected.

Appendix H - The Gravitational Potential at the surface of a star

The force of gravity on a mass m at a distance r away from a star of mass M *is equal to:*

$$F = \frac{GMm}{r^2}$$

The work done in pulling the mass from the surface of the star out to infinity is therefore:

$$W = \int_R^\infty \frac{GMm}{r^2} \, dr$$

This works out to be

$$W = \frac{GMm}{r}$$

and therefore the gravitational potential difference $\Delta\phi$ between the surface and ∞ is

$$\Delta\phi = \frac{W}{m} = \frac{GM}{R}$$

Now by definition, the gravitational potential at ∞ is zero so the potential at the surface is negative, hence:

$$ph \quad -\frac{GM}{R}$$

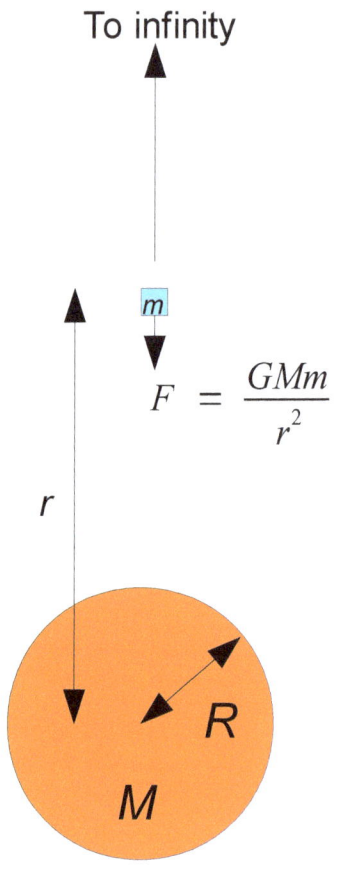

To infinity

$$F = \frac{GMm}{r^2}$$

r

R

M

www.ingramcontent.com/pod-product-compliance
Lightning Source LLC
Chambersburg PA
CBHW041102180526
45172CB00001B/78